HUMAN ERROR IN PROCESS PLANT DESIGN AND OPERATIONS

A Practitioner's Guide

HUMAN ERROR IN PROCESS PLANT DESIGN AND OPERATIONS

A Practitioner's Guide

J. Robert Taylor

CRC Press is an imprint of the
Taylor & Francis Group, an **informa** business

CRC Press
Taylor & Francis Group
6000 Broken Sound Parkway NW, Suite 300
Boca Raton, FL 33487-2742

First issued in hardback 2019

ISBN-13: 978-1-4987-3885-9 (hbk)

Library of Congress Cataloging-in-Publication Data

Taylor, J. R. (Consultant)
Human error in process plant design and operations : a practitioner's guide / J. Robert Taylor.
pages cm
Includes bibliographical references and index.
ISBN 978-1-4987-3885-9
1. Chemical plants--Safety measures. 2. Chemical plants--Design and construction. 3. Errors. 4. Human-machine systems. I. Title.

TP155.5.T388 2016
660--dc23 2015020198

Visit the Taylor & Francis Web site at
http://www.taylorandfrancis.com

and the CRC Press Web site at
http://www.crcpress.com

For Hanne Schøler Taylor, companion, friend,
scribe, editor, gentle critic and wife through
many years and many adventures

Contents

Preface

The first attempts at writing this book were made in 1978 when I was working under Professor Jens Rasmussen. My own interests were in design error, and I had had ample opportunity to see many design errors while working as a young engineer in the nuclear power industry.

Professor Rasmussen led a very active group, working on reliability and risk assessment, especially on the human reliability aspects of it. I had the opportunity to see the skills–rules–knowledge model take form and to provide a practical way of applying it in risk analysis. This led to the action error–analysis method, which has served for many years well as a tool for process plant safety engineering, and to an extensive validation project which demonstrated the value of human error analysis in industrial application.

The main problem that remained at the end of the 1970s was that many types of human error could be identified and even predicted, but the likelihood of them occurring could not be determined. Considering that some types of error are very rare but that it is the rare ones which are often the most serious, knowing the likelihood of errors becomes important. It is easy to drown in the hundreds and thousands of possibilities.

During the 1980s it became clear that laboratory research could not provide sufficient information of the actual conditions under which errors arise in industry. In any science, at some stage, observation is necessary, and in the human sciences this means field observation. Unfortunately, very few companies are prepared to accept long-term, in-depth field studies without some return. I began to work as a consulting engineer, mostly in the chemical industry, and could obtain, by direct observation, a good insight into factors affecting the frequency of errors. It also became possible to collect very large amounts of accident and near-miss data.

The one problem in this was that it never seemed possible to find the time, until now, to gather all the information together and present it properly. Luckily, I received some encouragement by several researchers and my younger colleagues' needs (there are some earlier publications; see Refs. [1–4]).

The main reason for writing the book at this time has been a practical need. Risk analysis for process plants, when following current best practice, has reached a stable and scientifically defensible state. Still, however, it does not address many of the realities of accidents in process plants. Human error is largely ignored in quantitative risk assessment for petroleum and chemical plants (in contrast to those for nuclear plants and aerospace systems). In a recent study [5], current risk-analysis methods, ignoring human error, were shown to be able to predict and calculate only about a third of the accidents happening in practice.

At the same time, designers struggle to develop plants which are robust and as error proof as possible. Information is needed in a form that they can recognise, and methods are needed which can be integrated closely into existing safety studies. Methods are especially needed which integrate human reliability analysis with hazard and operability analysis, which is the primary method of detailed safety study in the petroleum and chemical industries. Independent human reliability analyses are rarely successful, because the accidents which occur are never dependent on human error alone. More often they involve human error together with design weaknesses, management weaknesses and equipment failures. Methods are needed in which human error can be analysed alongside all the other contributors to accidents.

There are several good books which cover the theory of operator error (see the Bibliography). I have found them very interesting as background, but often they are quite lacking in close observation in the field. This is not surprising. It is difficult to study operators at work, and periods of observation of many months give only little feedback. The method of study here has been to work as an operator, as a commissioning coordinator and as an operation manager and to study plants along with the designers and operators in hazard and operability and human reliability analysis over a period of many years.

It is also the problem that while some books on human error discuss error assessment, actual predictive assessment is discussed only to a limited extent and validation of the techniques is almost nonexistent. While human error may be interesting in itself, for engineers its interest is primarily as a step to prevention or mitigation. If there is any need to understand the underlying psychology, it is in order to make prevention more effective.

This book was written primarily for risk analysts, safety engineers, designers and operators. Some operation supervisors and managers may find it interesting, but often, they know the content from their own experience. Cognitive scientists and human reliability specialists may find some new concepts, but the core methods described are traditional. Some of the practical aspects of application will be new. More to the point though, they should find the book useful as a source book for anecdotal and statistical data.

There has been a tendency in recent years to deprecate the term *human error*. The reasons for this are good. Errors are often the result of a complex web of influences. I can assure that human error does exist; I have made many errors myself during operating and managing process plants, luckily without serious accident. It does us no service to claim that the phenomenon is too complex to identify or calculate or that all errors are the result of management system failure. As will be seen, many, even most, errors are due to poor design, poor training, poor management systems, etc. Errors are caused, in many cases, by prior and external conditions. We need to understand these, as well as the internal causes, if we are to be able to make changes which will make work easier, more clearly understood, more congenial, and more error free.

The work described here is the result of a serial study over 35 years and does not take into account many developments in cognitive psychology since about 1980, so please be warned.

REFERENCES

1. J. R. Taylor, A Background to Risk Analysis, Risø National Laboratory, 1979.
2. J. R. Taylor, *Risk Analysis for Process Plant, Pipelines and Transport*, London: E & FN Spon, 1994.

3. J. R. Taylor, Using Cognitive Models to Make Plants Safer, in *Task, Errors and Mental Models*, ed. L. P. Goodstein, H. B. Andersen, and S. E. Olsen, London: Taylor & Francis, 1988.
4. J. R. Taylor, Incorporating Human Error Analysis into Process Plant Safety Analysis, *Chemical Engineering Transactions*, Vol. 31, pp. 301–306, 2013.
5. J. R. Taylor, Accuracy in Quantitative Risk Assessment? 13th International Symposium on Loss Prevention, Brugge, Belgium, 2010.

Acknowledgements

The prime debt owed in this book is obvious. Jens Rasmussen provided the theoretical basis and much encouragement. The book covers the application of only a tiny part of Rasmussen's work, but the material here is based on his concepts of cognitive information processing, in particular the early work.

I would like to thank Morten Lind and Erik Hollnagel, with whom it was possible to discuss many aspects of the work in depth during the 1970s and early 1980s and from whom I learned much. The biggest debt, however, is to the by now hundreds of operators, supervisors and operation and maintenance managers who have discussed operations and management problems with me, at length and in breadth.

About the Author

J. Robert Taylor worked for a short time in nuclear power plant construction before moving to the United Kingdom's Atomic Energy Authority Laboratory in Harwell, where he worked in several areas including the reliability of computer control. In 1972 he moved to the Risø National Laboratory in Denmark, carrying out research on risk-analysis methods under Professor Jens Rasmussen. Notable efforts were validation of the hazard and operability study method (by helping to build a chemical plant, then helping to operate it), development of methods for automated hazard and operability analysis and early work on design error. The methods described in this book were developed at that time.

In 1984 Taylor moved to the field of engineering consultancy and since then has carried out risk analysis and safety engineering on most continents. Notable achievements include a comparative quantitative risk assessment for all the major hazard plants in Denmark, supervising risk assessments for most of the oil and gas installations in Venezuela and risk assessment for almost all the oil and gas installations in the United Arab Emirates. The risk-analysis methods Taylor developed have been used in space applications, assessment of automobile control system safety and assessment of weapon system safety.

Taylor is the author of the QRA Open software for quantitative risk assessment and the HazEX software for manual and semiautomated hazard and operability and human error analysis, all open systems. He also prepared the database Hazardous Materials Release and Accident Frequencies for Process Plant.

This is Taylor's third book on risk analysis and safety engineering topics. He now lives in Denmark, carrying out research on design error and the quality of risk analysis and occasional safety consultancy.

Acronyms

ALARP as low as reasonably practicable, meaning risk is reduced until its further reduction is either impractical or unreasonable in terms of the resources required
BLEVE boiling liquid expanding vapour explosion
CSB Chemical Safety Board
DCS distributed control system
EPA Environmental Protection Agency
ESD emergency shutdown
FEED front-end engineering and development
HAZID hazard identification
HAZOP hazard and operability
JSA job safety analysis
LOTO lockout–tag-out; a method of locking equipment so that it cannot be operated while people are exposed to its hazards
LTA less than adequate
MOPO manual (or matrix) of permitted operations
OSHA Occupational Safety and Health Administration
PSV pressure safety valve
PTW permit to work
QRA quantitative risk assessment
TRA task risk assessment
VTS vessel traffic services

1 Introduction

WHAT IS HUMAN ERROR?

The cause of error, like beauty, is in the eye of the beholder. An example can illustrate this.

CASE HISTORY 1.1 A Tank Overflow

On a fine early spring morning, operators in a refinery control room received an alarm from a storage-tank level instrument. It indicated a high level in the tank and a potential for overflow. There were two level sensors on the tank, one using the radar principle and the other using a float, a wire (Figure 1.1) and a counterweight. The radar sensor gave the high-level alarm; the float chamber sensor indicated that everything was OK. The radar sensor had failed several times earlier over the previous year; the float chamber sensor had never been known to fail. The operators made no immediate response; the problem was noted to be checked on the next inspection round, and a work-order request was made for maintenance on the radar sensor.

In fact, the tank level was rising and the tank overflowed. The overflow was detected by a field operator, by smell. In all, 20,000 litres of fuel oil was lost. The overflow in principle could have caused a fire, but no ignition occurred. This might have been pure luck, but the tank had a ring ditch around it and a drain to an oil separator, so the oil did not spread much and the vapour cloud produced was limited.

One could well say that in this case the operators made an error of judgement; they relied on a balance of probabilities, not on a balance of risks. Shutting down the production unit was possible, but such shutdowns typically require a whole day of downtime, because distillation units take some time to restart. They could have put, though, the unit on 'full reflux', so that the oil just kept going round in the distillation column, while the problem was investigated. A switch to full reflux would cost about $30,000 per hour while the incident was investigated. The overflow actually cost much more than this in lost product and could have cost very much more if the oil had ignited, while a shutdown costs about $3 million.

This places the error clearly in the judgement of the operators. The error is understandable enough, based on a very strong expectation that the alarm was false.

FIGURE 1.1 The sticking level gauge wire. Operators did not believe it had failed because of long experience of reliable performance on many tanks.

In today's climate of expectation and regulation though, the buck does not stop there. There was no standing operating procedure which stated that all potential hazard alarms must be investigated and an action taken to prevent accidents while the alarm condition is present. This is today regarded as a management error, although often this view is expressed only after an accident has occurred. This makes two errors, one of judgement on the part of the operators and one of lack of judgement, or possibly, lack of awareness, on the part of the managers.

The story goes even further. One would expect operators to know how much product they have in a particular tank. Most operators producing to tankage *do* know such things. They know the production rate (this is a key measure of how well they are 'driving' the plant), and they usually know when the tank was last emptied. In addition, often the degree of filling is tracked on an operator's log.

Also, it is usual to record the level in the tank by computer, and present-day control systems (distributed control systems) should provide a record of the level variation over time, that is, a trend curve. Looking at this allows the flat part of the filling curve to be shown where the float chamber level gauge is stuck. No such record was available in this case, which today would be regarded as a design oversight.

So in all, there were five errors:

- A judgement error of the seriousness of the problem by the operators
- A management error of not having standard operating procedures with use of checklists
- An operative weakness in not knowing the tank content
- A management error in not requiring tracking of production
- A design error in not providing trend recording of the tank level

The example shows a pattern which will be repeated in many accidents. The occurrence required:

- A technical failure
- An operator error
- Omission of setting of operation standards by management, or lack of enforcement of existing operations standards
- A design weakness or error

It must be said, that as little as 20 years ago, only the first error on this list would have been regarded as such. Errors are defined by the deviation from what we expect as proper behaviour. When our expectation of proper behaviour changes, so, too, the way we allocate cause to the incidents. Often, the error will be defined as a departure from procedures and regulations. (This is still the case in most authority investigations of accidents.)

Over the period of just over 20 years, our expectations of proper operational behaviour have risen. Now for every accident there should be prevention. And behind every human error, there is another human error that caused it or failed to prevent it. The only accidents which do not involve human error at some level are what used to be called acts of god, but these could equally be called the unthinking cruelty of nature. These are things like earthquakes, tsunamis, hurricanes, floods and landslides. Yet even here, our expectation is that designers and managers take measures so that such events do not result in industrial catastrophe. To see this change in expectations at the extreme, consider the disapprobation showered on the managers and company after the Fukushima Daiichi accident in 2011 [1]. The almost unanimous sentiment among the public and nuclear power specialists was that the company should have been prepared for the event.

How far back do we need to go in tracing human error causes? As far as we need to in order to achieve our objectives. As an example, for many of the accidents described in this book, the cause is traced to lack of knowledge on the part of operators or managers. This is because first of all, a large fraction of accidents *can* be traced back to omissions in training or education, but more importantly, this is an area where it is possible to make a direct impact on safety with relatively limited effort.

Root cause–analysis procedures [2] are intended to determine the full range of errors and weaknesses which contribute to an accident. Unfortunately, root cause analysis as it exists today is designed to be used only after an incident or accident.

Is there any way in which all of the errors recorded above could be predicted? If there is, would we want to make a prediction, given that it probably would cost considerable engineering time? And if we did predict the errors, what would we do about them?

There have been several publications in recent years questioning whether the concept of human error can be defined [3]. In almost every case where an in-depth assessment of root causes is made, it is found that there is a network of causes and influences which is quite complex. It is hardly ever possible to attribute the error to the operator or the maintenance technician alone. The designer, the plant manager and others have almost always had a contribution to the sequence of events. That is why this book is called *Human Error in Process Plant Design and Operations*, and is not called *Operator Error*. When I use the term *operator error* in this book, the term is just a label for this network of causes and influences. It is convenient to do so, but we must remember that when we talk of human error, the question that must always be in our minds is, 'Who are the humans involved?' and we must remember that it may be an operator that pressed the button, but we will not get far in preventing accidents if we focus on the operator alone. We need to cast our net wide if we are to be able to predict errors and the resulting accidents.

CAN ERRORS BE PREDICTED?

Monitoring tank level is a primary activity in an oil terminal. It must be performed carefully and is well worth an in-depth study. For a crude oil distillation unit, product tank monitoring is just one out of several hundreds of operator activities. Nevertheless, the number of activities is not so overwhelming that we cannot study every one of them at a reasonable level of detail.

The case described in the previous section is an example of an error which occurs so frequently that it can be described as a syndrome. Operators frequently reinterpret an alarm as an instrument error. This is especially the case if the spurious alarms have been received earlier, if there are other indications which do not support the alarm or if the high level in the tank is 'inexplicable', as it would be if it arose from a leakage of oil or water into the tank through a 'closed' valve which is passing (leaking internally).

As will be seen later, errors of this type are very definitely predictable and can fairly readily be prevented. The only issues with this type of known error are how to identify the error opportunities *efficiently* and how to determine the level of risk, so that appropriate safeguards can be selected.

CASE HISTORY 1.2 Another Tank Overflow

An operator of an oil terminal checked that the tank was filling steadily, then left the control room to carry out valve lineup (opening and closing valves to create a desired flow path) at another terminal about 1 kilometre away. He stayed away longer than anticipated due to difficulties with the valves. The tank level rose. The high-level sensor was stuck and failed to stop the pumps so the tank overflowed.

This case can hardly be called an operator error. Having a single operator for field tasks and for control room supervision is asking for trouble, especially when the operator has responsibility for two sites at far-flung distances. Nevertheless, it could be said that the operator made a mistake, or several mistakes:

- Failing to predict the time to complete filling, or to judge the timing
- Trusting a single instrument, a high-level switch, as a single-point-of-failure dependency

The most critical error, however, is in a management which allowed for only one person to cover the tasks of field operator and panel operator and required him to work in such a large area. Erroneous operation was inevitable. Once again, the assessment of the causes of an incident depends on our expectations, and our expectations today are that managements provide a safe system of working. In most countries where I have worked, this principle is encoded in law.

In answer to the question in the heading of this section, it will be shown in the rest of this book that not only operator errors but also many related design and management errors *can* be predicted, and that in most cases, something can be done about them. The justification for human error analysis in an industrial setting lies entirely in the number of accidents which can be prevented.

THE IMPORTANCE OF OPERATOR AND MAINTENANCE ERROR

It is often said that human error is the cause of most plant accidents. Investigation of accident reports reveals that this is true; humans are involved as a cause or contributing factor in almost all accidents. The implication in the statement, though, is that most accidents are caused by operators. This is not true. There are far more groups of humans with a finger in the pie. Table 1.1 is taken from a study of the causes of nuclear power plant 'abnormal occurrences' made by the author in 1974, covering 14 nuclear plants over 10 years [4].

TABLE 1.1
Causes of Abnormal Occurrences (Near Misses and Minor Accidents) in 14 Nuclear Power Plants

Primary Error Cause	%	
Design error	35	
Oversight		17
Due to effect unknown to the designer		25
Dimensioning error		13
Complexity of the design problem		7
Communication problems		1
Unrecorded		22
Random hardware failure	18	
Operator error	12	
Omission of step in a procedure		49
Wrong procedure		16
Problem-recognition problem		7
Situation misrecognised		2
Sequence error		4
Operation on wrong object		4
Error in judgement of amount		2
Communication problem		4
Error in degree of adjustment		2
Maintenance error	12	
Fabrication error	1	
Error in written procedure	10	
Unknown	12	

As can be seen from the table, the designers contributed to a much greater extent than the operators to the near misses.

Figure 1.2 shows the distribution of cause contributions for 111 major hazard accidents reported to the European Commission [5].

As can be seen, management error or omissions and design errors have a far larger contribution to accident causality than operator and maintenance error combined.

This pattern continues today. A review of accident reports published by the U.S. Environmental Protection Agency and the U.S. Chemical Safety Board, which cover the majority of major accidents in the United States since the late 1990s shows that management and design error were involved in nearly all accidents (see Chapters 13 and 17). The topics of operator and maintenance error covered in this book obviously concern only a fraction of the human error problems in process plants. Nevertheless, operator error is an important problem, and methods to deal with it are needed.

More generally, it has been argued (1) that human beings are so complex that prediction is very uncertain and (2) that the concept of human error is relative, in that error is defined as deviation from *our* expectation. Incidence of error is then just as dependent on our expectation as it is on the person nominally committing the error. This is almost certainly true in general. In the field of process plant operation though, expectations, according to present law, should be well defined. There is a large grey area, in which managements fail to provide well-documented and practical operating procedures; where operators are encouraged to improvise in order to 'get things done'; where there is a gradual drift in the actual procedures followed in order to cut cycle times, increase production rates or reduce workforce and where managements habitually cut back on testing, inspection or maintenance. Fortunately, accidents arising from these causes are increasingly classified as arising from management error rather than operator or maintenance technician error.

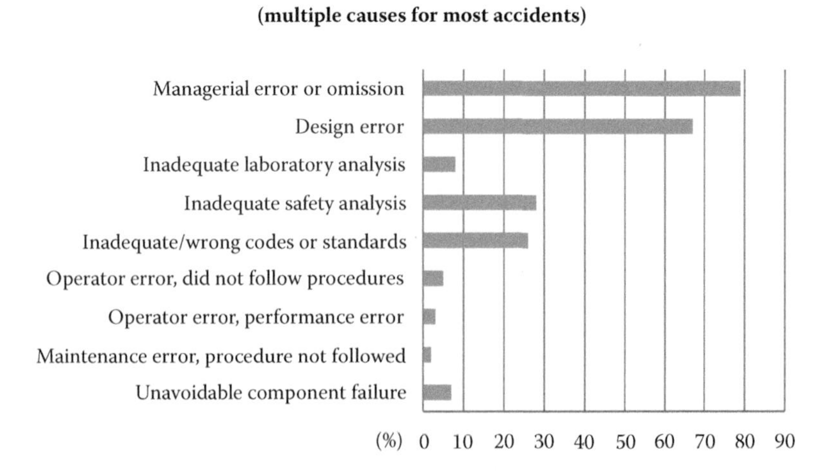

FIGURE 1.2 Distribution of error cause contributions. (Data from cases in Ref. [5].)

Remember, though, that managements are human and therefore prone to error. For our present purpose, it does not matter so much to whom the error is attributed, only whether the error can be predicted and prevented. And errors are defined relative to current best operation practice, for which there are many regulations and guides in the field of process plant operations. It is important though to investigate, and to be able to predict, the network of interactions and causes of error. If we do not do this, we may be able to supply a plaster on the sore for the most immediate, or proximal, cause of the accident, but there are many of these, and we can end in being covered in plasters. By understanding deeper causes and providing a cure, we can often avoid the need for plasters completely.

As will be seen in Chapter 19, it is possible to predict the frequency of many accidents arising from operator and maintenance error, within the strong confines of well-designed process plants with good operating practices and where best practice is at least a goal. There are still many plants around the world, though, where I would not care to make any predictions about where or how the next accident is going to occur, only that it will occur, and soon. Badly run oil, gas and chemical plants can be frightening, especially so for an experienced risk analyst who sees such plants as a collection of accidents waiting to happen. Accident prediction in such cases is a waste of time. It is much better to spend time on bringing a plant up to the minimum of current standards and requirements of safety practice.

COMPLEX SYSTEMS THEORY

Rasmussen [6] pointed out that operators in a process plant are generally required to push the limits of operation, to seek boundaries in order to extract the best performance of the plant. An operator error is then a consequence of a perfectly normal attempt to optimise, which unfortunately results in exceedance of the limits. If the operator action results in more production, the operator is praised. If it results in an accident, the operator is blamed.

Complex systems are those in which there are many acting influences, many feedback systems and many different limits, most of which are known to a few only, and some of which will only come as surprises.

Perrow developed the theory of 'normal accidents' [7]. He posited that modern systems are so complex that any local optimisation would tend to throw a system to its limits, and the systems could be so sensitive that small changes could result in a catastrophe. Dekker [8] and Woods et al. [9] have amplified that theory.

There are many systems which are definitely complex in this sense. Electrical power systems, telephone systems, many air traffic systems and the Internet are of this kind. (There have been many cases where millions have been unable to withdraw money from banks due to a few lines of coding error.)

Systems of this kind which are subject to high sensitivity and unpredictable catastrophe are called 'brittle'. The systems need to be operated in ways which are definitely suboptimal, in order to provide resilience. By the definition of complex systems, it is impossible to predict all the things which can go wrong.

The oil, gas and chemical plants considered here are generally complicated, but they are not complex in the sense described above. A large refinery has about 600 items of equipment, defined at the levels of pumps, vessels and distillation columns. These interact with each other in very well-defined ways through items called pipes. Each item of equipment has about 4 to 10 physical effects important for its functioning and typically about 20 things which can go wrong. At a deeper level, there are components attached to the equipments, such as instruments, and these interact by means of items called cables. Each has 1 or 2 effects involved in functioning and maybe up to 10 involved in failure. Overall, there are about 200 physical effects which need to be understood. It can be said that the systems are truly complicated, but this does not mean that they are unpredictable. Once the mechanisms are known, they are generally quite simple.

Over the past 40 years or so, methods have been developed so that not only the normal operation but also the disturbances and failures can be predicted (see, e.g. Ref. [10]). To a very large extent these systems are controllable, even in failure. Influences can occur across the well-defined boundaries of vessels and piping, such as those resulting from jet fires or explosions. Even these are predicted and as far as possible prevented, or mitigated to have only local effects, in a well-managed system, that is. There are cases where interactions are allowed to fall into chaos, as in those presented in Chapter 13.

To test the degree to which the above statement is true, we can revisit the accidents and near misses which have occurred in plants subjected to hazard and operability (HAZOP) and risk analysis by the author over the last 40 years, covering 93 analyses and 371 plant units. In these, 32 major hazard accidents occurred in all. Of these 32, 2 were not predictable and 1 additional accident was not predicted. The 2 which were not predictable involved chemical reactions which we cannot reproduce in the laboratory even yet. The third was a chemical reaction which was overlooked but which could be found in school chemistry textbooks. The reader may well ask why the 30 accidents that had been predicted still occurred. In only 1 did the plant manager directly refuse to implement the recommendations made on the basis of the analyses. The refusal was paid for 6 years later in the form of 11 fatalities and a loss of $1.2 billion. For the others, there are many ways that managers can procrastinate in implementing good safety measures, of which the two most common are scheduling to make changes at the next major shutdown and requiring a detailed engineering study to determine the best way of handling a problem.

The limits of predictability of accidents in the chemical plant accidents can be illustrated by a counterexample. There are very few cases like the following one (four in the author's career), but they do illustrate that not everything can be foreseen.

CASE HISTORY 1.3 A Chemical Waste Plant

Some chemical plants that can be mentioned to be truly chaotic, and defying predictability, are community chemical waste plants. These are difficult to operate because it is so hard to predict what substances and problems the community will supply. As an example, in a shipment of waste paint cans packed in drums, we found over a period of 1 year mortar bombs left over from World War II, old silicate putty that generated hydrogen when left in contact with rainwater and lithium batteries, which have a nasty tendency to burst into flame when old and abandoned.

Chemical waste disposal plants need to be robust, and their procedures and their operators and managers need to be robust too. Usually considerable effort is put not only into reducing risk as such but also into reducing uncertainty.

In oil- and gas-processing plants, and in oil fields, all but one of the accidents proved straightforwardly predictable. The one which was unpredictable was also a chemical effect, absorption of small concentrations of hydrogen sulphide onto a drying agent, which then releases concentrated toxic gas when wetted by rain.

In all the four cases of 'unpredictable surprises' found in follow-up of 93 risk analyses referred to above covered over 12,000 predicted scenarios for 4700 unit years of operation [11]. The most convincing reason to believe that modern oil, gas and chemical plants are complicated, rather than complex, is that automated hazard-identification techniques (automated HAZOP analysis, action error mode analysis and automated sneak path analysis) have been able to predict all accidents from the large set of historically recorded accidents and also a fairly large set of new accidents occurring over a period of 10 years [12]. Again, the unpredicted accidents were those due to hitherto unknown chemistry.

What can be seen from this is that unpredictable accidents are very rare, provided that one is prepared to make use of the full range of analysis techniques.

PUSHING THE LIMITS

As described above, one of the things which makes a system complex and brittle is the need to push the operations to the safety limits. There is no doubt that such effects happen and that they can be a cause of accidents.

CASE HISTORY 1.4 Driving a Plant Too Hard

Crude oil strippers are used to remove light gases, such as methane, ethane, propane and butane, from crude oil so that it can be stored safely in unpressurised tanks. During the first months of operation of a crude oil stripper unit, the plant was run at its extreme production rate, in order to determine how much oil could be stabilised and therefore how much the plant could produce. A stripper is a unit in which heat is applied in a column to drive off light components such as propane and butane from liquids such as oil, in order that the oil can be stored safely.

In this case the unit was operated with such a high flow of oil that the butane was not sufficiently separated. Unstabilised oil was sent to the storage tank, where the butane evaporated, lifting the floating tank roof. A cloud of butane vapour was released but fortunately did not ignite. The incident was classified as 'severe'.

This case is worth noting because it is very much the exception in modern plant operations. The incident is of a type known to operators as 'operation beyond the permitted limits'. Modern operation manuals have tables of permitted limits. In fact, the permitted limits are inside the safety shutdown limits, which are, in turn, within the design limits, which also are inside the safety limits (see Figure 1.3).

Of course, even with a clear guide in the operating manual, the operators could still push the limits, but in modern high-integrity operation (see Chapter 13) exceedance

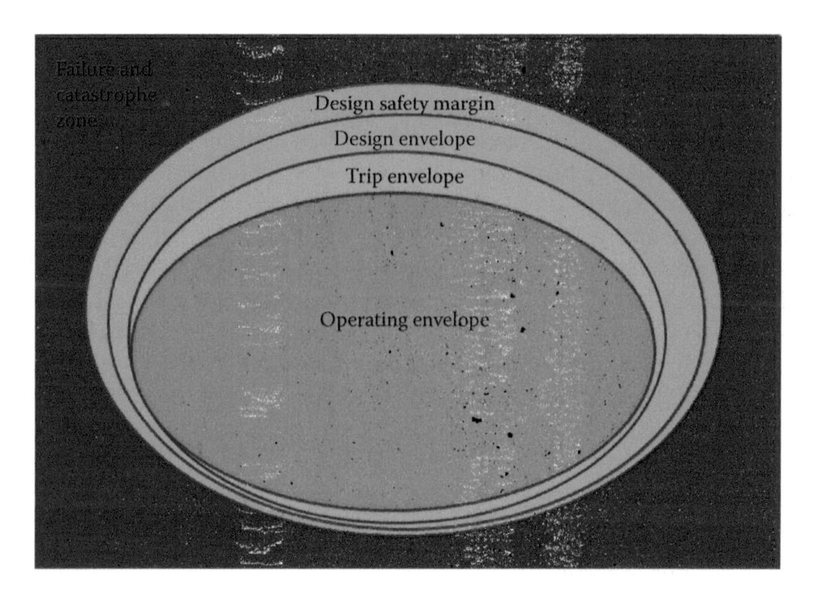

FIGURE 1.3 Operating envelopes for integrity in operation.

of operation limits is recorded by the control system. The number of exceedances is used to calculate a key performance index (KPI). This KPI, among others, is used to calculate year-end bonuses for everyone in the operating team. Not surprisingly, excessive experimenting with the plant limits is not encouraged—there is strong peer pressure to operate smoothly and without disturbance.

For the company, such 'good behaviour' is beneficial. Increased production by pressing the plant hard is quickly eaten up by one or two shutdowns arising from exceeding limits and causing safety systems to trip the plant.

This is what makes Case History 1.4 so unusual. It arose during a special period of testing the plant. This does not mean that there was no error in the case. Safety trips were bypassed, and there should have been a risk assessment which indicated the danger. The risk assessment should have resulted in special monitoring of the oil temperature and pressure, and there should have been special shutdown procedures which put limits on the testing. Special risk analyses for exceptional operations are now standard practice in the industry, including risk analyses for major (turn-round) maintenance, heavy lifts and special cleaning jobs.

Note that this level of safe behaviour, in which departures from approved practice are only made on the basis of risk assessments, is a feature of modern high-integrity organisations—it is far from the practice in many older plants; see Chapter 13.

SLIPS, LAPSES, MISTAKES AND VIOLATIONS

Reason [13] introduced some very useful terminology. He classified errors as follows:

- Slips and lapses—These are failures in execution or storage of an action sequence.

- Mistakes—The intention behind the action is wrong.
- Violations—The intention behind the action deliberately flouts safety regulations.

In the record of process plant actions, there are also cases of vandalism and sabotage (in U.S. Risk Management Plan 5-year accident histories, vandalism and sabotage were the causes of between 1% and 2% of incident causes in refineries and gas plants in the late 1990s; see Ref. [14]). The word *violation* seems a bit tame for such acts. Most violations, though, are less extreme, such as smoking in the plant, or driving too fast. Some defy belief.

CASE HISTORY 1.5 A Violation

Electrical cabinets were fitted with locks so that the lockout–tag-out procedure could be applied. No equipment could then be started while maintenance technicians were working on the pump motors. The circuit breakers were pulled out, then the door locked. An electrician found this very annoying, because his work was delayed by this. His solution was simple—he unscrewed the door hinges and removed the door, locks and all.

Case History 1.5 and some of the others described in Chapter 12 happened in plants with a very good safety culture. The first, Case History 12.1, put only the perpetrator himself in harm's way. Some of the violations could have killed others. We generally work in a 'no-blame' culture—it is more important to understand and thus prevent accidents than it is to apportion blame. In any case, there is no profit for the company in blame, and there is little effect in accident prevention by blaming. But in such severe violations, discipline is necessary; there is a need to send a message to all. For this reason, smoking in a refinery process or storage area is usually regarded as grounds for dismissal.

ASSESSING THE PROBABILITY OF OPERATOR ERRORS

The actual potential for error in Case History 1.1 is easily recognised: 'operator fails to respond to alarm' is a standard failure mode. 'Reinterpretation as instrument failure' is equally easy to recognise as a cause; it is a standard failure syndrome. Chapter 15 gives a list of such syndromes. We need only to organise these standardised causes in some way and then develop methods or tools to apply these to operator activities.

There are many tools which we can use to predict and evaluate operator error (see the Bibliography). With these tools, we can identify a very large range of possible errors, such as the ones described in the cases above. The problem which we then face is, 'So what?' We can find possibly 10 or 20 such potential errors in a simple task, such as filling a tank, and several thousand error possibilities in operating a process unit. Some will have serious consequences and will be reasonably likely, perhaps occurring once in the lifetime of the plant. Some will be extremely unlikely, perhaps occurring only once every thousand years. Without some way of determining how serious and how probable the errors are, we are left with just a pile of paper, recording more or less theoretical possibilities along with those of real concern and with no practical ways of setting improvement priorities.

Many error possibilities can be reduced to highly unlikely possibilities by following good practice in human–machine interface (HMI) design, plant design, operating procedure development, training, etc. However, from experience there will still be, in the end, large numbers of error possibilities for most reasonably sized plants.

It is doubtful that we can do anything effective to deal with long error lists containing thousands of possibilities. To be able to determine what action to take concerning potential error, we need to know how likely it is to occur and how serious are the consequences, or, in other words, to determine the risk. We need to be able to focus on the important errors. If the risk is high, we must do something, and using the as-low-as-reasonably-practicable (ALARP) principle, we can determine how much should be done by balancing the risk against the cost of accident prevention.

The probability of operator errors occurring in any particular task can be looked up in tables, notably in *Handbook of Human Reliability with Emphasis on Nuclear Power Plant Applications*, by Swain and Guttman [15]. (This is the 'old testament' of the human reliability field. There are several contenders for the role of the 'new testament', but my favourite is *A Guide to Practical Human Reliability Assessment*, by Kirwan [16].) There is also Table 19.1, which is based on observations over 36 years in 28 plants, from direct observations in the control room and in the field and from operation audits and incident investigations.

To be able to determine the frequency that an error is likely to occur, in addition to the error probability, we need to know how often the potential arises. This means, in Case History 1.1, that we need to know how often a tank will overfill, as well as the probability that the operator will ignore the alarm. The frequency of potential overfilling can be determined using standard failure-analysis techniques such as HAZOP or fault tree analysis, along with good tables of failure and plant disturbance statistics. Human error analysis and human error probabilities (HEPs) are needed to determine the frequency that human error will lead to overflow. Given that such overflow can lead to massive explosions, there is also a need to be able to estimate the probability of ignition. Safety equipment reliability analyses will be needed to determine the probability that alarm and trip systems will fail, and HEPs will be needed to determine the likelihood that operator response will fail.

For proper estimation of likelihood, we also need to take into account the factors which entered into the incident, such as an earlier history of false alarms, lack of clear standing orders for alarm investigation and lack of trend recording. In Swain and Guttman's book [15], such factors are taken into account by using 'performance shaping factors', of which they give about 20 such factors. For example, the complexity of a task is a performance-shaping factor. One of the problems I have found with this approach is that the received theory provides too few performance-shaping factors, and the standard mathematical approach to using them is seriously flawed. (This is discussed in Chapter 19.) Nevertheless, correct assignment of probabilities will need to take into account various causal factors.

The seriousness of consequences of an event can be determined by use of consequence analysis and simulation, or by referring to earlier accidents. For example, the consequences of the tank overflow could be simply the loss of product to an

American Petroleum Institute (API) separator (which separates oil from drain water) and the cost of cleanup, as in Case History 1.1. The consequences, under slightly different circumstances, could be an enormous explosion and the destruction of tanks and buildings, as happened in the Buncefield accident in England in 2005 [17].

Thanks especially to the U.S. Chemical Safety Board (http://www.CSB.gov) and the United Kingdom's Health and Safety Executive (http://www.HSE.uk.gov), there is by now fairly large collections of detailed incident reports available for major hazard accidents so that the degree of damage resulting from an error can often be determined directly from practical experience. Many of these are referenced in the Bibliography. A large collection of case histories is given in the books by Kletz, notably *What Went Wrong?* [18] and *An Engineer's View of Human Error* [19], and in the journal *Loss Prevention Bulletin* (http://www.icheme.org/lpb). A systematic collection of case histories indexed for easy retrieval is given in the systematic lessons learned analysis database in Ref. [21].

In the absence of good examples for a particular consequence type, the methods of consequence calculation used in quantitative risk assessment (QRA) can be used. In modern plant practice, a QRA will usually be available for a large process plant and can be drawn on to support human error risk assessment (and vice versa).

THE NEED FOR AN INTERDISCIPLINARY APPROACH

One of the things which becomes apparent from the examples discussed above is that human error risk analyses cannot be based on human error–analysis techniques alone. In order to determine whether risks are serious enough for it to be worth reducing human error, the frequency of error opportunities needs to be determined, the probability of failure of existing instrumented safety measures such as pump trips needs to be calculated and the consequences of the accident need to be calculated. For usefulness it is also necessary to know how to design for error prevention, reduction or mitigation. It is rare that good results can be obtained from a human error analysis in isolation.

This aspect of integrating human error analyses into overall safety and risk analysis has not been extensively described, although there are a few publications which focus on this [20]. Human reliability analysis needs to be a part of overall plant safety analysis, integrating with methods such as HAZOP analysis, safety integrity level (SIL) review and QRA.

The methods described in later chapters have all been used directly in large-scale risk analyses and HAZOP studies. Hopefully, the description will show how human reliability analysis can be made an effective part of the whole.

WHAT SHOULD WE DO ABOUT ERROR POTENTIAL?

The examples described above show that there can be many possible errors in a process plant and many possible causes for each. The solution to doing something about them, in Case History 1.1 specifically, could be any of the following:

- Providing standing orders to check explicitly every mismatch between two critical instrument readings, such as those of two level sensors, irrespective of the cause

- Providing a special alarm in the control system, to be given when two critical instruments have a mismatch
- Providing trend curve displays, so that the operator can check the history of readings and see that measurements have failed
- Providing more reliable instruments
- Investigating all false alarms and making an appropriate response
- Providing adequate staffing, so that there is always one person in the control room and another one available for investigation
- Providing hazard awareness training, so that operators are better able to assess risk, rather than just probabilities
- Always putting the plant on 'pause' or 'recycle' when a severe problem occurs or is suspected of occurring and investigating the cause (for a crude oil distillation, this involves putting the unit on full reflux, that is, sending the product back to the unit for redistillation, rather than continuing producing)

In fact, all of these things could be done as a standard design and operating practice. Providing standing orders to cover the cases described is a normal activity in modern petroleum plant operations. Provision of reliable instrumentation should come from the hazard and operability and SIL studies which are required by standards and regulations. Provision of good displays is included in human factors or HMI guidelines.

Hazard awareness training has become more common in petroleum and chemical plants. A review of what is expected today in plant operation safety is given in Chapter 13.

So why should we need to analyse operator error if the problems can be solved by standard approaches which will be required anyway? Part of the answer lies in the last measure above for Case History 1.1. Putting the plant on recycle is expensive (although not as expensive as shutting down the plant would be). Operators will seldom ever put a plant on recycle just because they are unsure about an alarm reading. In many cases such as when a gas release alarm is received, the standard and mandated response is for someone to go and look to see if the alarm is real or spurious. The instruction given is usually 'shut down on *confirmed* gas release'.

Theoretically, the correct answer to the question of which safety measures to choose should be determined by risk analysis, so that the risk of tank overflow can be balanced against the costs of stopping the plant. The risk-based approach requires analysis of all of the technical failures, as well as the possible operator errors.

Another answer to the question is that the standing orders, the instrumentation design, the human–machine design and the choice of operator response all depend on analysis and prediction. It is easy to see what should have happened in Case Histories 1.1 and 1.2, after the event. The trick which needs to be turned is to be able to choose and specify in detail all of the measures needed to prevent human error, or to prevent the consequences, *before* the accident has happened. The analysis also needs to be credible and well argued, before practical engineers will believe in it and act on it.

DATA COLLECTION FOR THIS BOOK

Most of the ideas in Chapter 2 are illustrated by anecdotes, ambitiously referred to as case histories. Full case histories with in-depth consideration of root causes are available for most of them. However, as Dekker writes, 'The plural of anecdote is not evidence'. It is not true data either, because for any anecdote illustrating a theory, there is likely to be another anecdote illustrating an alternative theory. Nevertheless, anecdotes are useful in bringing a theory to life and making it credible.

There is considerable difficulty in obtaining good data concerning error in plant operation, because you need either to be at the site when the error occurs or to be in close contact with those making near-miss and accident reports. At the same time, researchers are not really welcome in oil, gas and chemical plants unless they are contributing.

The actual method used in deriving data for the book has been observations in the field. The information is collected in four databases:

- Direct observations
- The database of incidents and near misses, collected by recording observations during operation integrity audits and from incident and accident reports [19]
- The database of design errors, most of which were derived from design review and HAZOP analysis workshops; there are few better places to study engineers at work than from the position of a HAZOP analysis facilitator [20,21]
- The database of QRA hazards and recommendations, which records hazards identified, accidents occurring, recommendations made and recommendations followed up, covering just over 7000 plant unit years (units are used as a measure of plant size here in order to be able to compare large and small plants; a middle-sized refinery will have 10 to 15 production units). Follow-up of safety recommendations gives some insight into management safety attitudes and management error

These databases are published separately.

As a scientific effort, the data collection leaves much to be desired. The population of companies is self-selected, being those prepared to employ risk analysts and safety engineers. There have been a very few 'broken safety management' companies included in the data collection, but these have been ones definitely trying to achieve safety. No 'rogue' companies with poor safety standards and no safety intentions are included here.

In keeping with the no-blame culture encouraged here, no company names are mentioned, although the interested reader can follow up some of the cases via the References.

REFERENCES

1. World Nuclear Association, The Fukushima Accident, http://www.world-nuclear.org, updated 2015.
2. W. Johnson, The Management Oversight and Risk Tree, U.S. Atomic Energy Commission, 1973.
3. S. Dekker, *Safety Differently*, London: Taylor & Francis, 2015, especially pp. 60–61.
4. J. R. Taylor, A Study of Failure Causes Based on US Power Reactor Abnormal Occurrence Reports, *Proceedings of the Reliability of Nuclear Power Plants*, Vienna: International Atomic Energy Agency, Publications, Sales and Promotion Unit, 1975, pp. 119–130.
5. G. Drogaris, *Major Accident Reporting System: Lessons Learned from Accidents Notified*, Amsterdam: Elsevier, 1993.
6. J. Rasmussen, *The Human Data Processor as a System Component: Bits and Pieces of a Model*, Risø National Laboratory Report M-1722, 1974.
7. C. Perrow, *Normal Accidents: Living with High-Risk Technologies*, Princeton: Princeton University Press, 1984.
8. S. Dekker, *Drift into Failure*, Farnham, UK: Ashgate, 2011.
9. D. D. Woods, S. Dekker, R. Cook, L. Johannesen, and N. Sarter, *Behind Human Error*, Farnham, UK: Ashgate, 2010.
10. J. R. Taylor, *Risk Analysis for Process Plant, Pipelines and Transport*, London: E & FN Spon, 1994.
11. J. R. Taylor, Does Quantitative Risk Analysis Help to Prevent Accidents?, in *15th Int. Symp. Loss Prevention and Safety Promotion*, February 2015.
12. J. R. Taylor, Automated HAZOP Revisited, http://www.itsa.dk, 2014.
13. J. Reason, *Human Error*, Cambridge, UK: Cambridge University Press, 1990.
14. J. R. Taylor, Hazardous Materials Release Frequencies, http://www.itsa.dk, 2007.
15. A. D. Swain and H. E. Guttman, *Handbook of Human Reliability Analysis with Emphasis on Nuclear Power Plant Applications*, Sandia National Laboratories, 1983, published as NUREG/CR-1278, U.S. Nuclear Regulatory Commission.
16. B. Kirwan, A Guide to Practical Human Reliability Assessment, London: Taylor & Francis, 1994.
17. U.K. Health and Safety Executive, *The Buncefield Incident 11 December 2005: The Final Report of the Major Incident Investigation Board*, Surrey: The Office of Public Sector Information, 2008.
18. T. Kletz, *What Went Wrong?: Case Histories of Process Plant Disasters and How They Could Have Been Avoided*, Fifth edition, Amsterdam: Elsevier, 2009.
19. T. Kletz, *An Engineer's View of Human Error*, Institution of Chemical Engineers, 1991.
20. J. R. Taylor, Lessons Learned from Forty Years of HAZOP, *Loss Prevention Bulletin*, No. 227, pp. 20–27, Oct. 2012.
21. J. R. Taylor, Systematic Lessons Learned Analysis, http://www.itsa.dk, 2014.

2 Models of Operator Error

When looked at en masse, accident reports involving human error appear as a confusing, amorphous mass. Each one seems unique, or at best similar to one or two others. I have often heard plant superintendents and operations supervisors say, 'You cannot analyse human error; there are too many possibilities'.

The thing that makes it possible to look at accident reports, extract the error descriptions, classify them and investigate each possibility is a model of error.

Over the years, many such models have been developed [1]. This book focusses on just one model, the skills–rules–knowledge (SRK) model developed by Jens Rasmussen in the 1970s. There are two reasons for choosing this model. Firstly, many of the more recent models have been based directly or indirectly on this, or have developed some aspect of it. Secondly, and more importantly for this book, it was this model that led to the development of the analysis methods described here. Most importantly of all, the data presented here have been collected systematically over a period of 36 years with a basis on this model and have been categorised according to the model concepts.

Models of human error are rather mechanistic descriptions of how persons process incoming information, how they react and what they do in response. More advanced models also describe a person's goals and objectives and the decisions and actions taken to achieve those goals. For operator error, the operator model must be linked to a corresponding model of the plant, in a way similar to that of the control system.

While it may seem somewhat inhumane to regard operators as a 'component' of the plant, it is even more inhumane to demand that they should perform perfectly all the time while knowing that every other component in the plant very definitely has a failure rate. We must accept that operators have certain environmental requirements, that tasks must be adapted to their capabilities and that even under the best circumstances, errors will occur. The best that we should expect is that in a well-designed plant, and with well-trained operators, errors will be rare and that dangerous errors will be extremely rare.

As will be seen, operator error is a significant contribution to safety-related incidents. It tends to be a prevalent aspect of the more serious incidents. Also, operator intervention is an important factor in achieving safety, and the reliability and speed of response in preventing accidents becomes important.

The SRK or 'stepladder' model of operator error was developed by Jens Rasmussen at Risø National Laboratory over the years 1971 to 1974, by which time it was largely complete in structure. There have been modifications and extensions since, by Jens Rasmussen himself, by his team and others. The extensions which were found to be needed to explain observations, however, were those needed to explain communications between operators, the dynamics of the teamwork and the impact of management error. These aspects were discussed by Rasmussen et al. [2] and have led to much more extensive treatments of cognitive processes, but these have been beyond the scope of this book.

At the time that the SRK model was being developed, the detailed mechanisms of neurophysiology were beginning to be investigated by neurophysiologists around the world, both by the established technique of reviewing the posttrauma psychology of brain trauma patients and by invasive studies of signal processing in the brains of cats and apes. For control engineers it was obvious that at some stage, the working of very large parts of the brain would be explained as a linkage of a fairly large number of clusters of neurons functioning as pattern-recognition devices. Morten Lind, a member of the Rasmussen Risø team, investigated the power of perceptrons (forerunners of today's neural networks) as a paradigm for human control [3]. It was natural to think of mental functioning as taking place in a network of associative processors. Today, the general form of these models has been confirmed by means of brain-scanning techniques to the extent that even the centre (or one of the centres) for empathy, ethical decision making and moral conscience has been localised.

Synthesising psychological studies, neurophysiological studies and advances in learning control, with observations of behaviour (and error) in the control room, Rasmussen wrote *The Human Data Processor as a System Component: Bits and Pieces of a Model* [4; 1974], and 'Accident Anatomy and Prevention' [5; 1974]. The use of the models in practical human reliability studies was described in *Notes on Human Factors Problems in Process Plant Reliability and Safety Prediction* [6; 1976]. It proved possible by the late 1980s to make a quite good simulation of human error potential so that error prediction could be made for some errors (semi)automatically [7] with quite good reproduction of error phenomena. One of the most convincing aspects of this work was that we were able to predict some kinds of error which only later were found in accident reports.

The SRK model is illustrated in Figure 2.1. The following is a rather mechanistic description of the model, deliberately so, because the mechanistic aspects will be used in the following section.

At the lowest left of the model is a 'processor' for first activation, then observation, which takes information from the human sensors (primarily eyes and ears, in the present context) and processes them into classified signals.

The classified signals are passed upwards to an identification processor, which determines what the signal represents.

The next step is to recognise the signal, in terms of standard procedures and practices in the plant, and to select the procedure to be carried out in response.

Having found the procedure to be executed, the different phased and timed signals are passed to the 'motoric controller', which, in turn, sends coordinated signals to the muscles, so that action is taken.

So far, this description is a classic control theory model, with relevance, though, to operator actions, since the different stages can be observed in recordings of operator responses to disturbances. What make the SRK model special are the information flow paths which bypass this full sequence of steps.

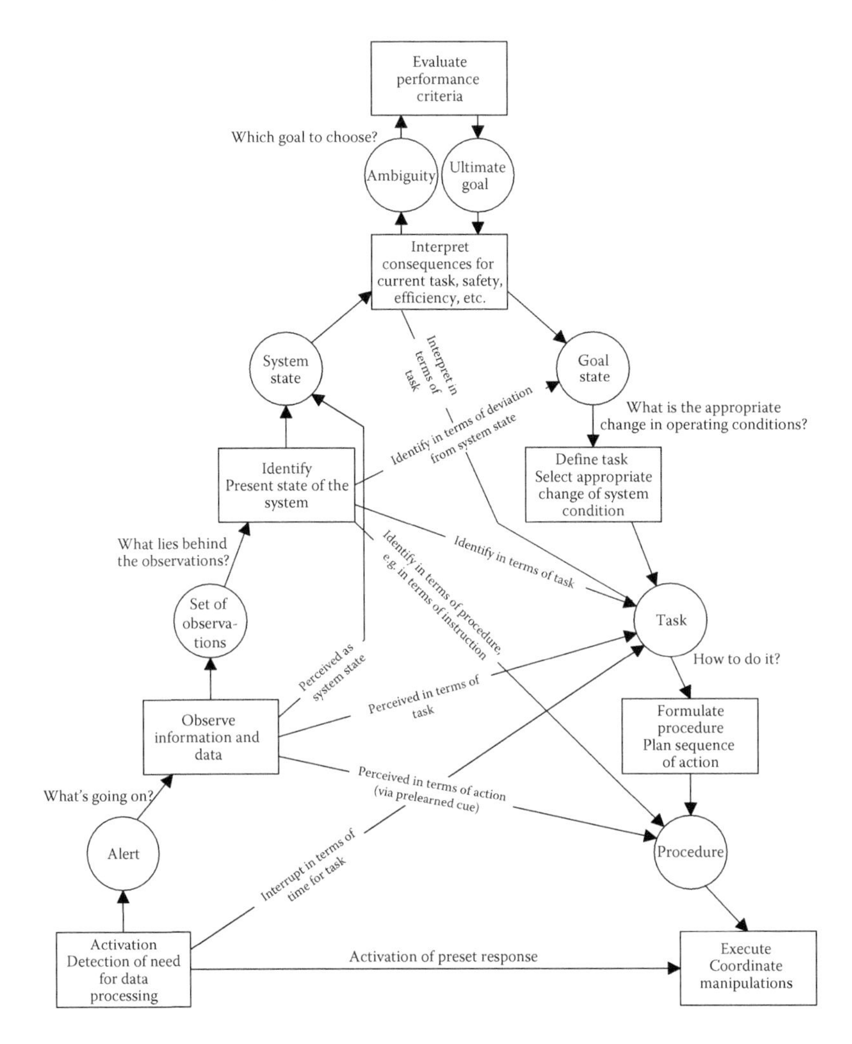

FIGURE 2.1 Stepladder model of the sequence of human processing activities between initiation of the response and the manual activity. (Adapted from Rasmussen, J., *The Human Data Processor as a System Component: Bits and Pieces of a Model*, Risø National Laboratory Report M-1722, 1974, by permission.)

If a set of observations is recognised as a standard known situation, a known required response sequence may be selected. This is forwarded to the procedural processor, which will organise the response and execute the steps.

THE SKILLS–RULES–KNOWLEDGE MODEL

The SRK model as originally formulated was regarded with optimism, because it allowed several error types to be predicted, which were later identified in accident reports. One of the most significant is 'capture due to limited observation'—passing straight to a standard response because a situation is 'identified' without even observing all the available information. Several accidents involving errors of this type were identified, including the key error involved in the Three Mile Island nuclear reactor core meltdown (see Table 9.1). The ability to predict error types which had not yet occurred at the time when the model was developed is one of the strongest arguments for its use. There have been many subsequent improvements in the model, both by Rasmussen and by others, but the original shown in Figure 2.1 is retained here because it has proved to be the most fruitful in studying human error in the field. Figure 2.2 gives a simplified version.

The usefulness of this model lies in the fact that it can be subjected to failure mode and effects analysis just like any other control system (still regarding the operator as a part of a control system). It is these error modes which form the basis for the error-assessment methods described in the remaining chapters of this book.

All of these models could take the form of rules on what happens in equipment X as a result of event Y. A rule-based simulator of this process which could imitate operator actions and errors with a good degree of fidelity was built during the 1980s; see Ref. [7].

One of the benefits of the model is that it can show how errors depend on knowledge. The operator needs knowledge of the rules (the operating manual instructions), the actual structure and equipment in the plant, the working of the

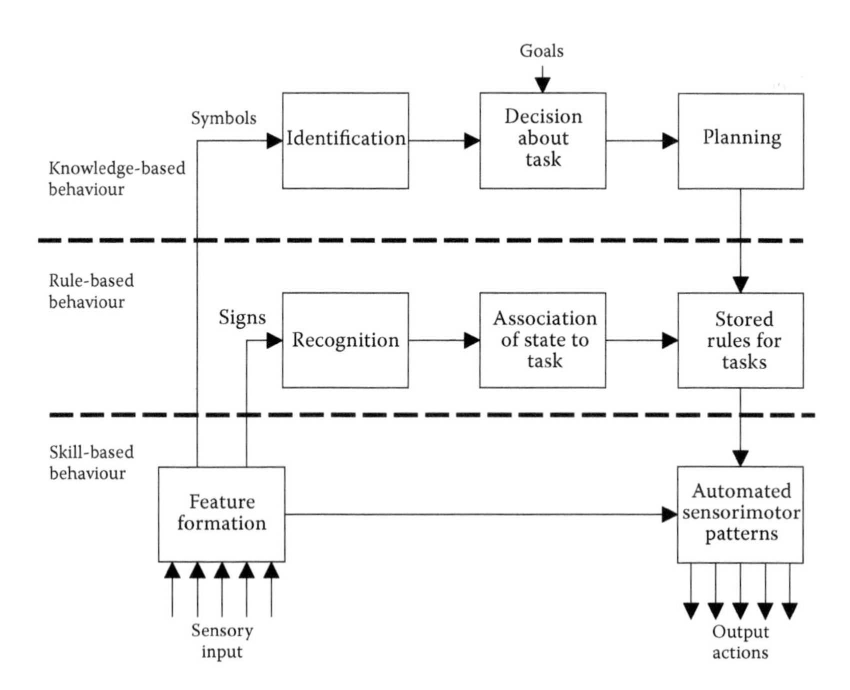

FIGURE 2.2 SRK model in its simplified form for event analysis. (With permission from J. Rasmussen, *Event Reports as a Source of Human Reliability Data*, Report N-23-79 Risø National Laboratory Denmark, 1979.)

equipment and general knowledge of the physics involved in plant performance. As we shall see, the operator needs not only mental models of how things work but also models of how things fail. The models will need to be both general models ('how do pumps work?') and specific models ('how fast will number 10 tank fill?').

There will often be a need for extensive sequential and possibly conscious thinking, e.g. 'I know the pump gives 600 litres per minute; the tank is only half full, so it will be several hours before I need to worry'.

Recent studies have shown that there is a significant lag between the actual decision making and the realisation which we call conscious thought, to the extent that some have considered that there may be no free conscious thought at all. When viewed as an automaton, this does not seem to be a serious issue; we have several busy processors in the brain, observing, taking decisions and doing things. And we have a top level which is called consciousness, aware of itself, watching at least some of the subordinate activities and believing it is doing all the work. The analogy to management of an organisation is too close to ignore.

SKILL-BASED ACTIVITIES

It is known that the knowledge-based reasoning and the rules part of the model does not explain all operator activity, because it would simply not be fast enough. Signals passing in neurons take about 300 milliseconds to pass from the eye, through processing, to the brain and to the hand under the best conditions. This does not leave much time for interactive mental data processing. I experienced this once, for example, in pulling a colleague away from a flexible hose which was beginning to whip. We calculated later, after overcoming the shock, that the action had to take well under one-third of a second; there was no time for conscious decision.

The requirements for riding a bicycle, catching a ball or adjusting the flow from a valve all require similar very fast response times. If you try to concentrate on catching a ball, thinking logically at each step how to move your body, hands and feet, you will drop the ball.

Speed is achieved in response by bypassing the rule- and knowledge-based control loops and passing signals directly from observation to the motoric systems. This is the skill-based part of the model.

The process of learning, in which repetition of a procedure leads to habituation, so that eventually what was a conscious sequence of acts becomes an instinctive whole, has been well researched [8].

RULE-BASED ACTIVITIES

When a new plant is handed over to the operators, an operator manual with detailed procedures is needed. The operators follow the procedures step by step. They do not necessarily need to know the reason for each step; they just need to follow the rules. Gradually the procedures will be learned, and in a routinely running plant, the operators will know the procedures and rarely, if ever, refer to the written form.

KNOWLEDGE-BASED PROCESSING

The processing up to this point is the rules part of the SRK model. If no clear situation and corresponding action rule is identified for response to the observations, a signal, with some detail or summary, of the observation should be sent to the diagnosis processor.

The diagnosis processor applies recognition rules (perhaps several rules, sequentially) until the situation is diagnosed. A signal is then sent to the decision processor, which combines the information from the diagnosis with general performance objectives, to develop a goal for further action. A signal to imitate this is sent to the planning processor, which applies pattern recognition to develop a sequence of actions to be taken (or possibly, a single signal for a stored procedure) to the procedural processor. This, in turn, activates the motoric controller.

The diagnostic, the decision-making and the planning processor will all need a model of how the plant works (or more generally, how the world works). One of the diagnostic strategies is certainly (from observations; see later) to hypothesize a cause of the symptoms observed, then to mentally simulate what effects would ensue. This includes observable effects which confirm or disconfirm the hypothesis. Similarly, the decision-making processor and planning processor need a model of the plant, in order to determine answers to questions such as, 'what would happen if … ?'

STATUS OF THE SKILLS–RULES–KNOWLEDGE MODEL

The SRK model is to a large extent a logically necessary one, to explain how operators adjust flows, carry out routine operations without knowing their full theoretical basis and solve control problems which they have never seen or heard of before. Parts of the model are confirmed by the predictions made using the model and from advances made in the study of neurophysiology, particularly from brain-scanning techniques.

What was not known at the outset was whether there are more levels, more processors, etc., which actually play a role in the information processing in the brain. From the current stage of development of neurophysiology and of cognitive science, it certainly appears that many areas of the brain and many types of information processing are involved. The 'true' model for operator action is presumably much more complex than that described by the SRK model as such. For present purposes, this does not matter. The model allows a very wide range of errors to be identified and to be described after the fact and to be predicted.

It is obvious also that there must be some short- and medium-term memory associated with some of the processors, because operators can ask for more information, or can look for it. This implies that a partial or preliminary decision may be made, or a conclusion reached, which is confirmed at a second stage by seeking confirmatory information (this mechanism also is known to lead to new error types in its own right, which provides some confirmation of the principle).

It is also clear that there must be feedback from actions and conclusions; otherwise, learning would not occur.

There is much more research to be made in order to map the model onto the functional centres of the brain, and there are probably many such mappings which are applicable to different persons. Until the time when this can be done, the SRK model is justified by the way in which it can predict errors and help in error prevention.

USING THE MODEL

Having turned the operator into a collection of interacting associative automata, we can now use it to predict error types. (Later, we will turn the operator back into a human being—there are many decidedly human aspects of human error.)

We can apply the techniques of failure mode and effect analysis to the above model, and ask the question, 'What errors would we expect if the model were accurate?' For example, what would happen if there were two very similar rules and the operator's observations were uncertain. One would expect the rules to be confused at some stage.

An example of this is an accident type which I have had to investigate twice and which has occurred many times around the world. A pipeline operator sees on the control panel that the pressure at the end of the pipeline is falling and responds by increasing pumping pressure. This is appropriate (in some cases) if the problem causing the pressure drop is blockage or high viscosity of the liquid in the line. It is decidedly inappropriate if the cause of the fall in pressure is a large leakage or rupture in the line. (See Ref. [9] for a tragic example of such an incident.)

In the following chapters, the different types of (posited) mental processing will be described and the ways in which errors can arise, illustrated by examples.

REFERENCES

1. E. Hollnagel, *Cognitive Analysis and Reliability Method*, Amsterdam: Elsevier, 1998.
2. J. Rasmussen, L. Goodstein, and A. M. Pejtersen, *Cognitive Systems Engineering*, Hoboken, NJ: Wiley, 1994.
3. M. Lind, *Investigation of a Class of Self-Organising Control Systems*, Risø National Laboratory Report R-315, 1976.
4. Rasmussen, *The Human Data Processor as a System Component: Bits and Pieces of a Model*, Risø National Laboratory Report M-1722, 1974.
5. J. Rasmussen, Accident Anatomy and Prevention, Risø National Laboratory, 1974.
6. J. Rasmussen and J. R. Taylor, *Notes on Human Factors Problems in Process Plant Reliability and Safety Prediction*, Risø National Laboratory Report M-1894, 1976.
7. J. R. Taylor, Using Cognitive Models to Make Plants Safer: Experimental and Practical Studies, in *Tasks, Errors and Mental Models*, ed. L. P. Goodstein, H. B. Andersen, and S. E. Olsen, London: Taylor & Francis, 1988.
8. L. Argote, Organisational learning, Springer 2013.
9. A. B. de Souza Jr., E. L. La Rovere, S. B. Schaffel, and J. B. Mariano, Contingency Planning for Oil Spill Accidents in Brazil, Freshwater Spills Symposium 2002, 2002.

3 Hindrances and Inability to Function

Many omissions of operations occur because the operator is simply not present or unable to react. A list of typical reasons for this is the following:

- Operator absent from the workplace
- Operator hindered or obstructed
- Incapacitation
- Comfort call
- Distraction
- Attention failure
- Action prevented or hindered
- Overload
- Priority error
- Focus error

OPERATOR ABSENCE

One cause of operational input being overlooked is that of the operator simply being absent. Case History 1.2 in Chapter 1 is of this type. There are others.

CASE HISTORY 3.1 Absence from the Work Location

A tank truck driver started filling his tank truck with gasoline. Since it would take a long time, he left the filling and went to the smoking room. He stayed too long, and the tank truck overflowed. The gasoline caught fire and the loading racks burnt down.

CASE HISTORY 3.2 Absence from the Work Location—Another Case

An operator was to drain water to an open drain from a waste oil tank. The draining would take some time, so he went to the smoking room after opening the valve. By the time he returned, about 5 cubic metres of oil had drained to the water-treatment plant, allowing oil to overflow to the fjord and requiring a complete plant shutdown.

CASE HISTORY 3.3 Yet Another Absence from the Work Location

A ship traffic controller in a vessel traffic services centre was absent from the observation radar in order to clean a few cups. During this time, a ship with the mate asleep at the helm missed a waypoint and sailed into a bridge. About 50% of the bridge width was lost. Ironically, the vessel traffic services centre had been designed in part on the basis of the mate falling asleep at the helm at just the waypoint concerned. Even more ironically, the automatic collision-prediction function of the radar, which would have prevented the accident, had not been implemented because of cost and because the probability of overlooking the collision potential on the radar had been considered to be low.

In plants with a few personnel, with just one board operator, the operator may be absent for a short period to make coffee or tea or visit the toilet.

Error-Reduction Measures for Absence from Work Location

Long operations in which the operator is required to stand and wait should be avoided. If such steps are necessary, a rest area should be provided in which the operator can sit and still observe the operation. In warm countries the location should be shaded, and in cold countries shelter should be provided from cold. Steps should be taken to ensure that the task is interesting, or some kind of alarm should be provided to arouse or call the operator.

HINDRANCE

Operators are sometimes hindered from performing their job by physical obstruction. Locked doors, materials set down in access ways and materials not removed due to poor housekeeping are all examples. In one case a supervisor checking completion of instrumentation work before giving permission for start-up was unable to check some items because scaffolding had not been removed and was hiding the equipment. Even worse, the ladders *had* been removed so that he could not climb to make the inspection.

In some cases it is difficult or even impossible for the operator to react to a situation because of the physical hindrances built into the plant. One example is that of a valve located high on a plant with no access ladder or stairs.

More frequently in accidents it is fire, smoke or explosion that prevents an operator responding.

CASE HISTORY 3.4 Prevention of Action by Fire

In the Piper Alpha accident of 1988 [1], operators were not able to coordinate the response because the control room was damaged by the explosion. Also, personnel could not activate standby fire pumps because of the fire. Two persons who put on protective clothing and breathing apparatuses and went to start the pumps were never seen again.

Obstructions due to poor storage and housekeeping can prevent operator response in an emergency.

CASE HISTORY 3.5 Work Hindered by Laying down of Materials

In an audit survey on a large gas plant, five of the fire water monitors which would be needed in case of fire were obstructed by piles of materials stored and made ready for the next major maintenance.

INCAPACITATION

Operators may be unable to undertake a task or respond to an alarm because they are dead or ill. Equally, they may be asleep or intoxicated.

Death is one form of incapacitation and has quite a high probability when compared with other causes of an operator failing to function. In the relevant age group and job class, the probability of fatality is about .025 per year (in the countries studied). This gives a failure frequency of about 2.5×10^{-6} per hour, which is comparable with many high-quality plant components. It corresponds to a failure rate of 5×10^{-3} per year while actually working, which is significant when compared with other error frequencies (see Chapter 19). No good data could be obtained for the frequency of collapse due to illness in the appropriate groups, but from health records and unsystematic observation, the frequency of incapacitating illness seems to be two to three times higher than the fatality rate due to nonwork causes.

Error-Reduction Measures for Incapacitation

The possibility of death or illness is one of the primary arguments for the provision of dual staffing for a lightly loaded control room job. The buddy system is a similar requirement in many companies for operations in the field.

Another possibility is the provision of devices similar to the dead man's handle used in rail and public transport, which ensure a fail-safe action unless the person is able to respond. A similar concept is to provide check-in points, with swipe cards, for persons making plant inspection tours. One of the most effective recent innovations is the provision of wireless monitoring of operator and other personnel location. Personnel-monitoring systems can also be provided with some vital signs monitoring. While these are generally provided for ensuring operator safety and for rescue in accident situations, they also provide a degree of security for the plant as well.

For the problem of intoxication there are many measures for limiting its probability, which in some cases is required by law. These measures include strict prohibition of intoxicating substances and in some cases periodic or random blood or urine tests.

DISTRACTION

Distraction is one of the causes of failing to detect signals or observations, which should lead to operator intervention. Distraction can take many forms.

Talking on the telephone represents a form of distraction.

CASE HISTORY 3.6 Distraction by the Boss

An operator was discussing with his supervisor the need to replace a control valve. It would require a complete plant shutdown unless the valve could be bypassed and the flow controlled manually. The discussion was serious because the shutdown would mean that production targets could not be met, but manual control would be difficult and would require an additional operator to be brought in.

During this discussion, a high-level alarm occurred on one of the distillation columns for which the operator was responsible. It took some time for the operator to notice the alarm and to disengage from the discussion. By this time the lower trays in the distillation column were flooded, causing some damage. The plant shut down at a high level, but the shutdown was a little too late to avoid damage because of a set point error.

Other forms of distraction are discussions taking place in the control room, visitors and incidents taking place in other parts of the plant and dealt with by other operators.

Error-Reduction Measures for Distraction

A culture which allows the operator to focus on the job and discourages interruptions is necessary for safety. Supervisors must avoid involving panel operators in lengthy discussions, as should others. When control panel activity is low, operators need to be able to talk to others or to do something else such as filling out the operators log or reading the newspaper. It is preferable, though, to design the operator's job so that there is enough happening to maintain interest. If this cannot be done, all upset conditions must be provided with audible and visual alarms.

ATTENTION FAILURE

Concerns away from the job (money, family worries, ambitions or just the weekend fishing trip) are often cited as the reason for distraction. That this does occur is certain. I have observed operators with a 'six-mile stare', completely out of contact with the task in hand. This occurs especially when there is very little happening, when the plant is operating steadily and presents nothing to think about. From long observations in control rooms, such distraction seems to occur only rarely. Nevertheless, it is sufficiently important that control room designers and operations managers should design the control job so that there is a sufficient task load to keep the operator involved. Also, displays should be made in such a way that the operator can be continuously aware of the actual plant status and has a routine requirement for supervision of the processes.

Control panels and workstation displays are designed to attract attention with audible alarms and flashing lights when attention is needed so that the effect of distraction is minimised. Distraction becomes very important for the case of persons such as crane operators and their banksmen (helpers), who have a very varying task load and few electronic aids to help remaining focussed.

Error-Reduction Measures for Attention Failure

Operator jobs should be designed to maintain a high level of interest sufficient to retain attention. If this cannot be done, all upset conditions must be provided with audible and visual alarms.

OVERLOAD AND PRIORITIES

Operators can be unable to respond to situations simply because there is too much work to be done. Situations with high-intensity operations arise when there are too few operators for the procedures to be carried out. It is not always the case that an operator can simply work through the workload. If vessels are being filled, prewarming is being carried out and heating is proceeding, the process itself may set the schedule. Even more so, in distribution systems, the demands of the users determine the workload.

In some cases, companies have determined the levels of staffing on the basis of normal workload. They plan that any special load, such as plant start-up, can be carried out by overtime working or by bringing in staff from other activities if this is necessary. This leaves operators in difficulties when unexpected additional work is imposed, for example, as a result of failures or process disturbances.

When there is too much work to do, it may appear obvious that the most important things should be done first. In fact, setting work priorities can be difficult. It may be difficult for an operator to foresee an impact or a delay and therefore continue what he is doing.

Alarm lists are not always arranged in suitable ways. One of the worst examples investigated in these studies involved a system where all alarms blinked on a mimic display until acknowledged by pressing an acknowledge button, at which stage the alarms went to steady red and the individual alarm texts were displayed *one at a time in order of arrival.*

Good modern human–machine interfaces have alarm priority displays. Priority must be set for each alarm, usually determined by the criticality of the related effect, e.g. high-level, temperature and pressure alarms having high priority. Sometimes priority is allocated on the basis of time to react.

Error-Reduction Measures for Overload

Reduction of errors arising from alarms has been extensively investigated, with results resulting in industry guidelines [2,3].

PRIORITY ERROR AND FOCUS ERROR

Straightforward errors can also occur in setting priorities, even when there is no overload. If operators are having difficulty writing a performance report, they will often have a problem in diverting attention from the writing to responding to an alarm, for example.

It is possible for operators to be so focussed on the work in hand that they cannot even see any other work. An example was that of an operator setting up a loading arm for transferring cryogenic liquefied petroleum gas to a ship. The job and the communication were difficult, so much so that the operator could not see a serious leak taking place only 10 metres away.

In many accidents, it is the noise which makes one aware. A large pressurised leak of gas is deafening, but even a small pressurised leak makes itself heard.

REFERENCES

1. Cullen, The Hon. Lord W. Douglas (1990). *The public inquiry into the Piper Alpha disaster.* London: H.M. Stationery Office.
2. Abnormal Situation Management Consortium, *Effective Operator Display Design*, Abnormal Situation Management Consortium, www.asmconsortium.com, 2008.
3. Abnormal Situation Management Consortium, *Effective Alarm Management Practices*, Abnormal Situation Management Consortium, www.asmconsortium.com, 2008.

4 Errors in Observation

The first processor in the SRK model of the operator is observation of input.

The failure modes identified by analysis of the model in Chapter 2 are:

- Input overlooked or not seen due to the following:
  - Input not visible
  - Input badly placed on display
  - Information hidden deep in a hierarchy of display screens
  - Input not part of operators habitual monitoring pattern
- Input too complex to process
- Ambiguous input
- Misrecognition of input
- Failure of alarms, indicators or annunciations

POTENTIAL OPERATIONAL INPUT OVERLOOKED

Input needed by the operator can be available in principle but not be visible to him. One classic example in a nuclear plant is that of an alarm which was mounted on the rear of a row of control cabinets. Another is that of an alarm lamp which was obscured by a hanging label on which was written a note that another alarm lamp was nonoperational.

In modern practice of control room design, workstations show a plant mimic diagram, usually with several levels of display, so that the operator can zoom in on a detailed presentation of a particular plant unit. Each operator typically has two display screens available, one which remains most of the time with the overall display, including alarm indications, and one which focuses on the most critical operating parameters. The operator can usually switch very rapidly between screens and can show alarm lists and trend curves as well as mimic diagrams.

One of the problems with this kind of arrangement is that much of the information needed by the operators is not continually displayed (Figure 4.1). They may switch quickly to the necessary display screen or 'page' but will only do this if they are performing a periodic browsing, are investigating a possibly minor disturbance or are responding to an annunciation or alarm. Alarms make the operators aware that there is something to investigate. If they are already in the process of responding to another alarm, the information may be effectively invisible. If there are many alarms, it becomes difficult to follow what is happening (see Chapter 21).

COMPLEX OR UNRECOGNISABLE INPUT

The complexity of inputs can cause operators to mistake equipment or misunderstand a situation. Their capacity to respond correctly depends on the degree of complexity and on experience with the actual equipment.

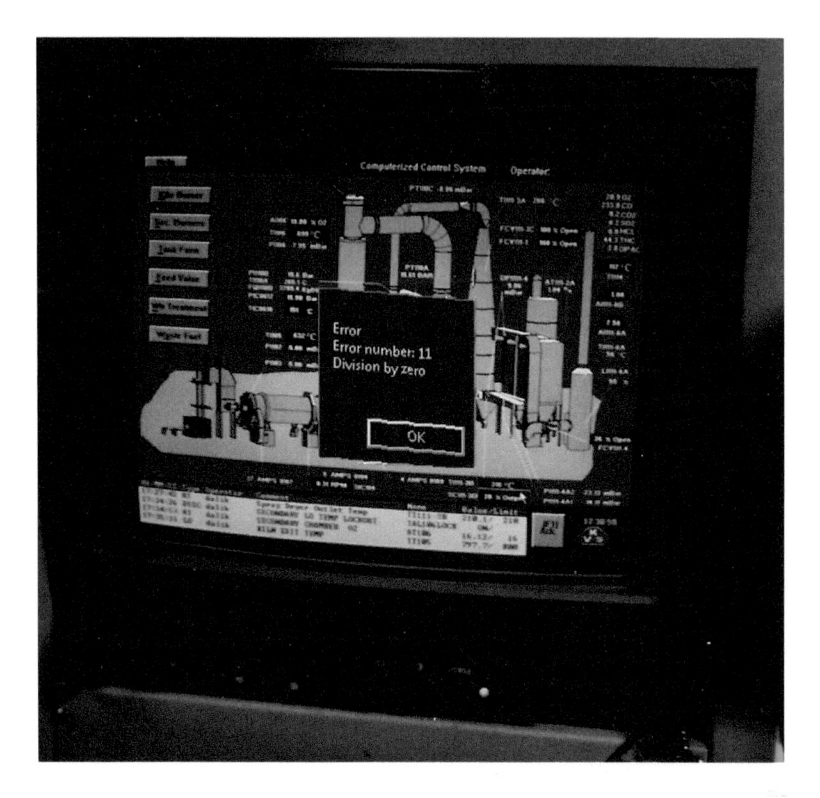

FIGURE 4.1 A modern take on the case of the alarm lamp covered by a label, in this case by a Windows™ dialogue box.

Error-Reduction Measures

Clear layout, good spacing and good labelling in the plant all contribute to good recognition. Clear and readable information on the control panel and the display screen contributes to good recognition in the control room, as do uncluttered display layout and standardised location of names, process variable displays, alarms and display screen navigation aids. Proper correspondence between control panel and plant labels is critical to proper understanding.

Names should be chosen carefully. 'Local' names for equipment should be avoided, but if there is a strong local naming tradition, designers should be made to follow it. They should not impose their own names onto equipment which are different from those widely used in the plant.

AMBIGUITY AND MISLEADING SIGNALS

Signals coming to the operator can be ambiguous, with the same indication arising in a case with several possible causes. An example is low pressure in a pipeline, which can arise from a failing pump or from a large leak. Another is a low temperature in a reactor, which can arise from a failure in the heating supply or can arise from

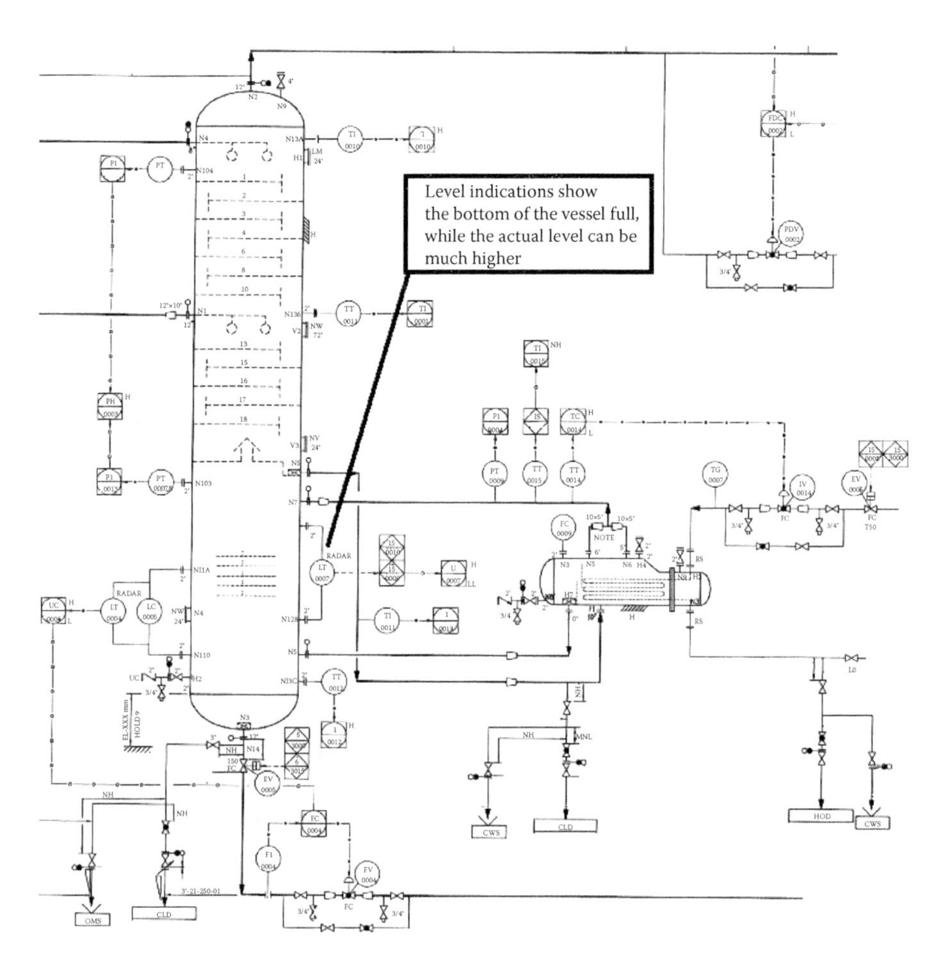

FIGURE 4.2 The board operator viewed this screen, which provided information on the raffinate product leaving the unit but not on the liquid being added to the unit. Also, the indication of level in the column was misleading. (Courtesy of U.S. Chemical Safety Board, Washington, DC.)

a lack of catalyst. In the second case, increasing the flow of heating steam can be catastrophic, because if the reaction ignites at a high temperature, the reaction may run away.

In the Texas City accident of 2005, the level indication of the distillation column indicated that the level was high, enough for liquid to reach the lower tray of the column, but did not show that the column had filled. The indication was directly misleading. Figure 4.2 shows a similar problem, common on many plants.

MISLEARNING

When deciding the importance of a process parameter, the operator needs to refer to a reference value. If this value is learned wrongly, the operation will be in error.

Alarms have also been involved in erroneous learning. If operators observe that an alarm has no particular significance, they may ignore it. If the operation is successful, this can then be built into the informal operating procedures. This happened at the Texas City accident of 2005, where operators were used to the level alarm for the distillation column being activated during plant start-up and ignored it.

FAILURES OF INSTRUMENTS AND ALARMS

Level-sensing instruments in particular have contributed to, or directly caused, many accidents. The Texas City Refinery accident in 2005 [1,2], the Milford Haven Refinery accident in 1994 [3] and the Three Mile Island nuclear power plant accident in 1979 [4] (Table 9.1) were all cases where stuck or limited level sensors caused extensive effects leading to major accidents but continued to indicate that at least the liquid level was OK.

REFERENCES

1. Chemical Safety Board, Investigation Report—Refinery Fire and Explosion and Fire. BP Texas City March 23, 2005.
2. The Baker Panel, "The Report of the BP U.S. Refineries Independent Safety Review Panel," 2007, Washington, DC.
3. UK HSE, The explosion and fires at the Texaco Refinery, Milford Haven. 24th July 1994, www.hse.gov.uk.
4. M. Rogovin, Three Mile Island: A report to the Commissioners and to the Public. Nuclear Regulatory Commission, Special Inquiry Group, 1980.

5 Errors in Performing Standard Procedures

Rule-Based Operation

Operator error is inevitable in systems of any design. The operators, like any other component in a process plant, have a certain capacity. When demands on the operators exceed their capacity, the result will be a failure or an error. The operators also inevitably have a stochastic aspect to their behaviour. If the system cannot tolerate the occasional randomness in their responses, the result will be an error.

The various places where operator and maintenance error can enter into plant failure behaviour are shown in Figure 5.1 [1].

During maintenance, testing or repair, an error can lead to an active disturbance in the plant or, more usually, to a latent failure or latent erroneous state. The latent failure can be revealed during normal operation and so leads to a disturbance.

An alternative pattern of disturbance starts with an erroneous operation under normal operation. Disturbances can also arise due to technical failures, and the disturbance brings the plant into an abnormal state. When a plant is brought into an abnormal state, normally there are possibilities of either automatic or manual safety actions; there is also the possibility of omission or incorrect execution, so that the correct safety result is not achieved or the disturbance is made worse.

Human error plays a role in performance of automatic safety systems. Errors in maintenance or repair can prevent safety action. Human error in testing can also mean that failures are not revealed. A possibility, which has been recognized as significant after the Three Mile Island incident, is that operators can interfere with safety system operation if they think that the safety action is dangerous.

Finally, there is the possibility of extraneous activities, unconnected with the plant or unconnected with operation, to interfere with performance. Examples are a fitter who drills a hole in a wall and then through instrument cables; or a welder who starts a fire with sparks; or an electrician who drops a spanner into a relay cabinet. All these can give rise to plant disturbances and arise from operations which are completely unrelated to the equipment or systems affected.

From this description, it can be seen that analysis of operation and maintenance error must cover a wide range of circumstances and types of error, even when the scope of study is restricted to standard procedures.

THE PROCESS OF RULE-BASED OPERATION

Carrying out procedures involves different thinking processes, which take place with different cues. Mostly, the initiator for an operational process is a decision in an operation meeting, a memo from production planning or a request.

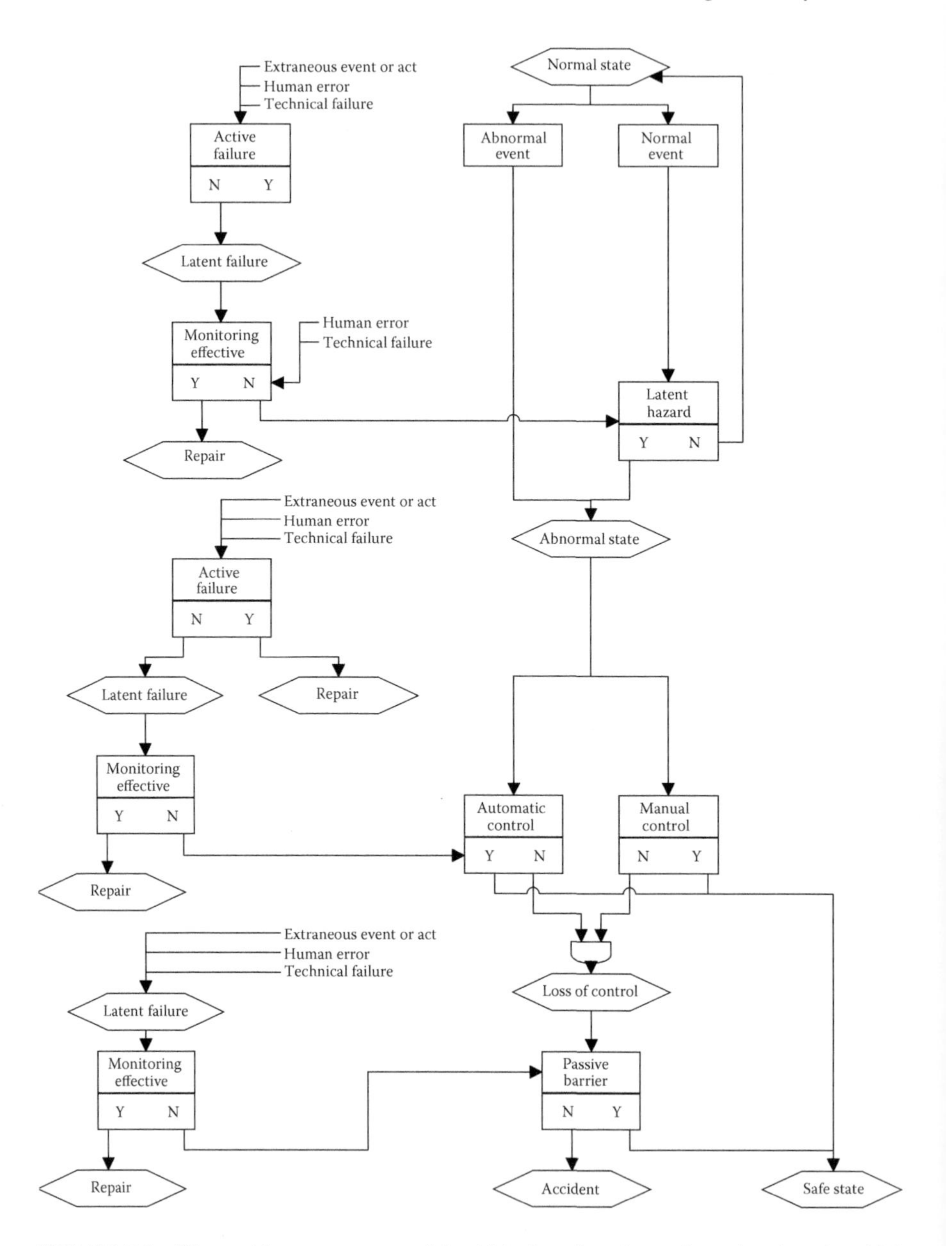

FIGURE 5.1 The accident anatomy model, which describes the various situations in which operator and maintenance can contribute to an accident. (Adapted from Rasmussen, J., Accident Anatomy and Prevention, Risø National Laboratory, 1974.)

The procedure may be one which is well known and therefore sequenced by the board operator or operation supervisor from memory. For complex procedures, with many equipment items, an operating procedure manual, or a structured checklist is more likely to be used in a high-integrity organisation. In some plants, written procedures are seldom used or even referred to, often because the written procedure is out of date or wrong.

In carrying out a procedure, the pattern of actions is typically as follows:

- Determine the next action to be performed (supervisor): This may be quite complex, involving several items of equipment. Actions may be cued, that is, by a signal from the system on completion of a process; or may be sequentially cued, where actions follow closely in sequence; or may be self-cued when the impetus for an action is generated by the operator, for example, when realising that 'the time is now'.
- Delegate the action to a board operator, to a field operator or to a board operator and the field operator team.
- Await a report for completion of the subtask.
- If a report that there are difficulties comes back, such as a valve failing to open make a decision on what is to be done and give new instructions.
- When the task is complete, confirm the effect if possible, then check it.
- Select the next action/task.

In a complex start-up, a supervisor may have one or two board operators active, each with responsibility for one or a few plant units, and may have up to two or three field operators active within each unit. The supervisor, therefore, has quite a lot to keep track of, especially in a large integrated plant or a power station.

The individual operators can have quite a lot to do also. Each action, such as opening a valve or starting a pump, requires checking that preconditions for the action are complete, actually carrying out the action and then confirming that the action was successful.

In older plants, confirmation of a successful action was often difficult. When starting a pump, for example, the board operator would ask field operators to open the suction valve, start the pump at a local switch panel, then open the discharge valve gradually. Of these actions, only the pump start would be visible at the control board, and correct pump operation could be determined only at the end of the sequence, via a flow indicator for larger pumps or a discharge pressure indicator, or in some cases only when level changes could be seen.

In modern plants, even ordinary block valves are quite likely to be automated in main process and utility flows, in order to minimise operation time. In these cases, the valves will generally have a switch indicating full closure/start of opening and a switch indicating full open condition/start of closing. The working of the valves can be confirmed therefore from the control room. For the pump start, there will generally be a suction filter delta P pressure indicator, to indicate whether the filter is blocking and/or a suction pressure indicator. For the discharge, there are likely to be both flow and pressure indicators. This allows not just performance but also correct effect of the procedure performance to be confirmed.

WHAT IS A PROCEDURE?

A rule-based procedure can be any sequence of tasks which is defined precisely beforehand. Table 5.1 gives an example of changeover of two valves to allow for valve removal for maintenance. The reader will note that the procedure is expressed very systematically and carefully. Nevertheless, there are many things that are left to the operators' knowledge. For example, the field operators are expected to know the position of the block valves and the location and method of operation of the vent valve. They are in particular expected to know how long to keep the vent valve open, which is simple enough if the block valves are close but can be difficult if one of the valves is located far away.

The names of the operator roles have been changed for consistency with other descriptions in this book, but otherwise the procedure text is an original procedure from a gas-processing plant. As can be seen, even so simple a task requires a good deal of care.

One thing that can be noted in the procedure is that there are several checks but no indications concerning what to do if the check result is unsatisfactory, such as in the case of gas leak in step 5, for example. The reaction would usually be to postpone maintenance on the second valve in case of gas leakage from the block valves or shut down the pressurisation of the plant in case that maintenance is essential.

THOUGHT PROCESSES IN PROCEDURE EXECUTION

There is a large difference in the way operators think about procedures. For some, a procedure is a collection of unrelated steps; for others, it involves creating flow routes and process conditions on a drawing or a control display graphic. For the most experienced, it involves flows in pipes and conditions in vessels which are visualised as such. From interviews (62 in all), operators with field operation experience tend to think of conditions inside the pipe and can often point to a pipe and tell the physical state, estimate the temperature and know what the flow rate should be. Others have little idea of these but know where the valves and pumps are and how to carry out instructions.

CASE HISTORY 5.1 Considered Thinking

One evening, in a control room, I had the opportunity, while writing up an operations log, of hearing our two best waste incinerator operators discussing the effects of blending of the waste (to obtain a degree of uniformity in combustion properties) on secondary combustion chamber temperature and the tendency for dioxin formation. Very little was happening in the plant, and they were engrossed for about half an hour, using the displays on one screen to follow their reasoning (while still displaying any alarms, which might come on a second screen). The discussion was of a high quality, with most of the relevant temperature/catalysis/concentration relationships discussed and related to feed blending difficulties.

None of our other operators, unfortunately, would have even understood most of what was discussed, although they did perform their job well. We consistently achieved lower dioxin concentration than most similar European and U.S. waste incinerators.

TABLE 5.1
Typical Checklist-Style Procedure

Title: Changeover of 001 – FV – 002 A/B (applies for change from A to B)
Purpose: Valve changeover to allow for maintenance
Warning: Gas is flammable and toxic.
Precautions: Ensure that there is no hot work proceeding in the area. Place powder fire extinguisher within 2 metres of the work location.
PPE: Use SCBA and personal gas alarm.

Action	Set Point	Clarifications, Checks, Cautions and Warnings
1. Initial state: B block valves closed.		If either block valve is open, close it.
2. Field operator strokes open and closes the B valve.		Field operator checks that the valve moves smoothly, from fully closed to fully open and back-checks with board operator that valve has opened and closed fully.
3. Field operator switches the B valve to manual and closes it.		Confirm the valve identity with the control room operator and confirm the initial valve positions.
4. Field operator opens the B block valves.		
5. Field operator tapes the B valve and blocks valve flanges and checks for leaks by using soapy water.		
6. Board operator slowly transfers the load from A valve to B valve.		Field operator monitors valve movement.
7. Board operator switches B valve to auto.		
8. Field operator closes A valve block valves.		
9. Field operator vents A valve to flare.		
10. Field operator loosens bolts on A valve and checks for gas release.		Loosen bolts slightly at first, and use flange spreader to ensure that gasket is not sticking and holding the flange closed.
11. Field operator removes bolts and turns isolation spectacle plate.		
12. Secure spectacle plate, insert new gaskets and bolt fast.		

Thinking of a process plant as a living entity, with fluid flowing in pipes, pumps pushing liquid, steam heat exchanges and levels rising in vessels as flow begins, provides a completely different way of checking procedures. Knowing what is happening 'within the steel' allows the operator to cross-check sounds, display screens and indicators to correct for omissions.

ERROR MODES IN CARRYING OUT PROCEDURES

Errors in the second area of the SRK ladder operation activity are confined (by definition) to inaccurate execution of well-learned procedures, described as sequences of 'actions'.

Possibilities here include the following:

- Omission of an operation, a step or a subprocedure
- Carrying out an action too early or too late
- Carrying out an action too quickly or too slowly
- Carrying out an action with too much or too little force
- Carrying out the task with too little precision
- Reading an instrument wrongly
- Skipping an action (which is not quite the same as omitting an action)
- Repeating an action
- Omitting to check a precondition for an action
- Carrying out a correct action on the wrong object
- Reading the wrong instrument
- Carrying out an action in the wrong sequence
- Carrying out an action in the wrong direction (turning an adjustment knob left instead of right, etc.)
- Carrying out the wrong action completely
- Combination of the above error types

In addition to these error modes, there are also possibilities of accidents by carrying out an operation correctly according to the operating procedure but with the plant in the wrong condition. An example is adding solvent to a reactor which is hot due to a valve leak in a steam-heating system, generating a massive release of flammable vapour as a result. Such latent hazard effects may be completely outside the range of what can be called an error, for example, if they are completely hidden or can be within the range of what one would expect an experienced operator to be aware of.

Figures 5.2 and 5.3 show examples of the application of the error mode key words.

FIGURE 5.2 Operation too skewed; load may fall off.

FIGURE 5.3 Operation too fast; liquid is sloshing.

CAUSES OF ERRORS IN CARRYING OUT PROCEDURES—FOLLOWING CUES

The main group of supervisory control errors is failure to respond to a need for action. There are several groups of causes. The first are action-cueing problems:

- The human–machine interface (HMI) does not indicate the system state due to error or omission in the HMI design.
- The cues for action are hidden on a control cabinet panel which is not visible from the operator position.
- An alarm or annunciator is covered by a hanging label tag, report, newspaper, etc.
- The annunciator lamp has failed.
- The instrument needed for an action cue has failed.

- A necessary cue is simply not displayed.
- The operator does not know the significance of a cue.
- The operator is busy with some other task, telephone call or another activity.
- An operator is busy with a high-priority task.
- The operator cancels the audible alarm and then forgets to process the alarm.
- A computer-based mimic diagram display is too cluttered to allow the operator to determine the action or action priorities.
- A computer-based alarm display is too small to allow all alarms to be displayed and priorities are not properly allocated.
- There are simply too many alarms to allow a proper response to all of them (often as a result of a cascade of disturbances).
- The operator mistakes the identity of the cue.

CAUSES OF ERRORS IN CARRYING OUT PROCEDURES—OMISSION

By far the most common error in executing procedures is omission, especially the steps which are unconnected with the main purpose of the task, such as restoration of valve lineup after a test or refilling a tank after maintenance. In many cases this error appears, from interviews after near misses, to be caused by distraction. Since the task itself is completed, there are no more necessary steps to achieve the goal and, as a result, nothing to remind the operators if they are distracted.

Other causes of omission of steps in procedure which have been seen in incidents are as follows:

- The cue for starting the procedure is missed, as described above.
- Items in a long list are to be checked, but one or more items have not been added to the list. This has happened relatively frequently when new subsystems or improvements have been installed.
- Items are not checked when operators work from memory. Memory may be faulty or simply deficient because facts have never been learned (inadequate training).
- Operators transferring from another unit may know, in general, how to operate, but small differences can lead to errors or omissions.
- Distractions may be a cause of erroneous operation, as they are for observation errors. The operators simply forget where they are in the procedure.
- In complex, multiperson procedures which last over several shifts, distraction can extend to such a degree that a lead operator or supervisor simply loses track of what has and what has not been done. (This is one of the prime reasons for using tick-box checklists.)
- At shift change, lack of communication about what has been done and, more importantly, not done can lead to omissions.
- Simple forgetting can be a reason for omission.

A group of problems in this area is trapping, in which a particular cue pattern is definite, but part of the cue pattern is ambiguous. If the operator uses only part of the cue, and responds to this, then the response can be in error.

The more complex disturbances in a plant will often not have a direct planned response, so that there is no standard procedure. In this case, diagnosis will be required. This kind of activity is treated in Chapter 9.

Sometimes indications from the plant are ambiguous. At some point this will always be true. A high pressure in a distillation column feed tank can, for example, result from an outlet valve blockage or closure, an inlet valve failure or a high feed temperature. In a highly instrumented plant, there will be sufficient instrumentation to disambiguate the problem. With less instrumentation, the board operator may need to ask a field operator to check valve positions.

Even if there are sufficient inputs to clearly indicate a problem, it will not necessarily be the case that the operator makes the right conclusion. Operators often control the plant on the basis of a few key parameters, such as distillation column pressure drop, column overhead temperature and separator level. In this case the emphasis is on securing good performance when disturbances occur; the operator should look wider, but often will not.

Most of the operations carried out by operators are perfectly standard and can be carried out from memory. The operators have learned the cue to start the operations and the sequence to perform it. Generally such actions will be written in the plant operating manual for the plant units and in standing operating procedures for more generic operations, such as starting a pump.

Operators are generally able to carry out frequently performed procedures from memory. They may be required to follow written procedures whether they need to or not, in order that each step can be ticked off on a checklist. This usually applies only to items such as unit start-up and shutdown or a shift in operation configuration, which is carried out rarely (such as once per year), or where operations are particularly hazardous or critical for the product. In pharmaceutical production, use of checklists may be a requirement in order to provide documentation for authority approval.

For batch production, there will always be a sequence of operations, tasks and actions to perform. For modern plants, these will often be carried out by sequential control programmes, although these may be of two types. With true sequential control, actions are carried out automatically when the necessary time has elapsed or when necessary conditions are fulfilled. With operator-paced sequential control, the sequence of operations is fixed, but the start of each new step requires confirmation by the operator. Figure 5.4 shows the controller for such an operator-paced sequential controller. At each step in the process, a button lights, and the step is carried out when the operator presses the button.

For continuously operating plants, the cue to commence actions may be a request/order from the production manager, or from operators of other plants; an alarm or an observation of an adverse trend by the operator. Examples of these kinds of operations initiating cues are the following:

- Increasing trend lines, e.g. increasing level
- Alarms, e.g. a high-temperature alarm
- Messages from a field operator, e.g. there is a seal leak on a pump
- A fire and gas system alarm

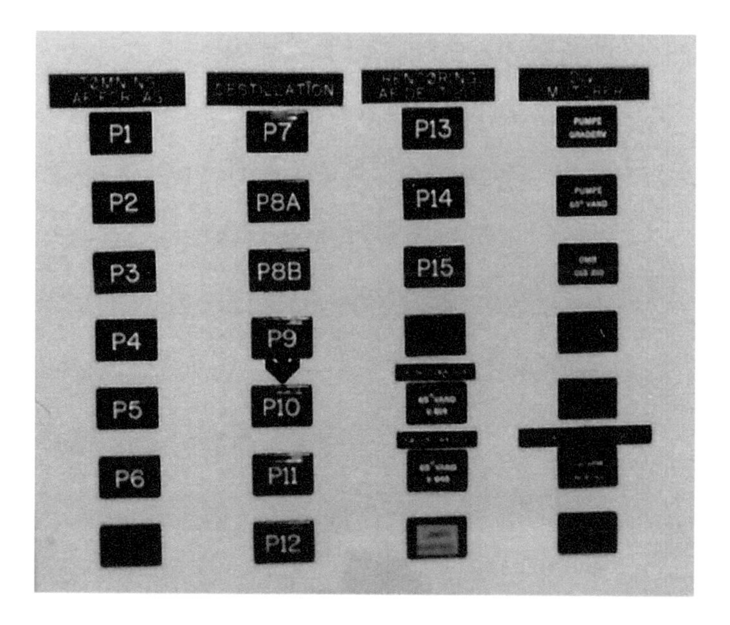

FIGURE 5.4 Operator-paced sequential control that uses a very early model of programmable logic controller. The correct button is backlit when the system is ready to proceed with the next production step.

These kinds of cues/responses are those of supervisory control and are discussed in detail in Chapter 8.

Other, much longer, sequences of actions are required during start-up, shutdown or changeover procedures.

ERRORS IN PROCEDURE EXECUTION—WRONG PROCEDURE

Operators may use the wrong procedure mostly for good reasons.

If there is no standard procedure but it is necessary to get a job done, then a procedure must be created. In the Texas City accident of 2005 [2], operators found the pressure in the distillation column rising. They opened a 3-inch blowdown valve, which had been noted as potentially dangerous due to the large blowdown rate. The reason was that the controlled depressurisation valve was not working.

There is nothing wrong with creativity in developing new procedures. This is necessary when situations which have not been foreseen by the design engineers arise. Such creativity is often needed in maintenance tasks, such as clearing blockages. However, no such tasks should be undertaken without at least a job safety analysis and where changing of valve lineups or process parameters is planned, a mini-hazard and operability study. Even in an emergency, this kind of analysis should be made, because it is too easy to make an incident into an accident.

CASE HISTORY 5.2 Procedure Adaptation

A 'drum shredder' was used to deal with drums of solid waste. The drums were fed into a hopper, with rotating knives, where they were cut into pieces, and the solid waste reduced while waste solvent was added as necessary to produce a 'paste'. The mixture was then pushed by a high-pressure concrete pump to an incinerator.

A problem arose due to some barrels already containing liquid, hidden beneath solids. In this case the mixture would become too wet. Pumping the solid part of the waste was still possible, but the liquid would run back through the solid, until a large liquid content built up, which could dangerously spill into the incinerator.

It was suggested that the solution to the problem would be to simply add more drums of solid waste; on analysis this was determined to be hazardous as it could lead to mixing incompatible solids or overloading the shredder. Instead, a system was developed for adding dry sand to the mix, which was confirmed to be safer.

An insidious problem in executing procedures is 'trapping'. If a situation looks like a standard one, then the standard procedural response is performed. The result can be catastrophic if the actual situation is not as expected.

CASE HISTORY 5.3 Procedural Trapping

Operators were transferring gasoline from a refinery to a remote storage terminal. The flow rate at the terminal fell. This was not unusual; changes in pump performance, temperature or product viscosity could cause this (although not to any great extent with gasoline, but the operators did not know that). They responded in the usual way by increasing pumping rate. Unfortunately, the reason for the low flow rate at the terminal was caused by a leak in the pipeline. About 20,000 barrels of gasoline were lost. Fortunately, there was no ignition, despite the release occurring at a high traffic intersection, and most of the gasoline could be pumped up.

Incidents like the above have occurred on a multiproduct pipeline transferring kerosene in Cubãtao, Brazil, in 1983 [3] (with hundreds of fatalities occurring when the kerosene flowed through a village and ignited), and at Ufa, Russia, in 1989 in a liquefied petroleum gas pipeline.

Traps such as these can be readily prevented by adding just a little more instrumentation—a flowmeter at each end of the pipeline and a pressure indicator for the pump discharge, together with proper emergency recognition hazard awareness and response training.

A 'perfect trap' is one in which the situation appears innocuous to the operator, when in fact it is dangerous or at least harmful. Other examples are given in Chapter 15.

CASE HISTORY 5.4 A Perfect Trap

One situation which occurred for me arose during commissioning of a distillation plant. Methanol solvent, distilled from the product, was to be returned to a ground tank. The tank had been installed, a manhole cover fitted and the pipe from the unit to the tank put in place. When pumping started, methanol was sprayed in all directions. The pipe had been put in place, but no hole had been cut into the manhole cover, and the pipe did not enter the tank. The pipe fitting had been done so well that it was impossible to see the difference between properly installed pipe and an incomplete installation.

Mislearning or erroneous teaching is a cause of errors in selecting the appropriate procedure. Frequently, accident reports refer to inadequate training. (See Chapter 7.)

Mislearning can also occur when operators move from one plant to another, but when their experience is not so directly transferable.

CASE HISTORY 5.5 A Normal Procedure Is Dangerously Inadequate

Repairs were to be made on an acid tank. The tank was drained down and washed, then ventilated. Before allowing entry to the tank, the atmosphere inside was tested for breathable oxygen and for flammable atmosphere, both at the entry manhole, and at the top vent. The tank was signed off for work. When welders started welding about 1 hour later, an explosion occurred. Hydrogen had collected under an internal box intended to prevent inflowing acid from disturbing settlement. Because the box was inverted, the light hydrogen was prevented from being swept out of the tank during ventilation. Because the box was inaccessible, the gas inside could have been tested only after tank entry and with use of a ladder, which did not occur to the gas tester.

CASE HISTORY 5.6 Wrong Procedure Due to Instrument Failure

Instrument failures are a relatively frequent cause of performing the wrong procedure. At Three Mile Island in 1979 [4], a stuck level gauge caused operators to struggle to prevent overfilling of a pressurisation vessel, when in fact the problem was low cooling water level in the reactor vessel. In addition, safety valves had opened consistent with a high level. Actually a safety valve was stuck open and was the main cause of the low level.

Errors in emphasis in training were a factor in this accident too. The dangers of overfilling a reactor so that it could become hydraulically full, and possibly rupture, had been emphasised to a much greater extent in training than the dangers of low liquid level and meltdown had been.

CASE HISTORY 5.7 An Unfortunate Departure from Procedure

A kettle-type reactor producing phenylacetyl chloride reacted benzoic acid with thionyl chloride, a very corrosive and toxic compound of sulphur, oxygen and chlorine.

The reactor was filled with phenylacetic acid in sacks. The acid was then heated to melting and stirred. Thionyl chloride was added slowly because, although the reaction is endothermic and cannot run away, the reaction produces hydrogen chloride and sulphur dioxide. Too rapid a reaction would overpressure the system and overload the acid gas scrubbers.

In the actual incident, the phenylacetic acid was heated until 80°C was shown on the temperature indicator. In fact, the temperature was not so high in the bulk of the material. Solid caked residue had gathered around the temperature probe and the sensor was reacting to heat conducted from the heating coils. Instead of reacting, the thionyl chloride gathered at the bottom of the reactor. When, eventually, the mixture was heated sufficiently to react, the amount of thionyl chloride was large and reacted quickly. The pressure blew out the burst disk and hydrogen chloride and sulphur dioxide were vented to atmosphere.

ERRORS IN PROCEDURE EXECUTION—TOO MUCH OR TOO LITTLE

For any plant, companies should establish the safe limits of operation. In many countries, this is required by law. For pressure, upper limits are always specified. The limits are usually used to define set points for trips set points and for pressure safety valve settings. (Companies may decide to make set points below the safe operating limits, for operational reasons, reducing the frequency of unwanted trips.)

When plants are protected by instrumented shutdown systems, the task of the operator becomes one of controlling parameters as close as possible to optimal levels while not coming too close to automatic shutdown limits. Some skill is required in this, especially if there is significant variation in input parameters such as feed rates, product demand rates or raw material properties.

Where disturbances have relatively slow effects, as in most process plants, the operator will have alarms to attract attention.

In plants with rapid variation in parameters, continuous attention is required. For the most part, such high attention control is restricted to distribution systems with no storage to smooth disturbances. Examples are electrical power utilities and liquid distribution. Generally designers should provide sufficient storage or holdup capacity in the plant to ensure that responses in less than 1 minute are not required.

For some plants, and for some parameters, automatic shutdown systems are not provided. In these cases the operator is responsible for keeping the plant within safe limits. A typical example is heating of a batch reactor, where the operator often has to adjust the heating rate to compensate for varying heat release or to avoid too high temperatures at heat transfer surfaces. Error parameters on amounts can occur because of the operator's desire (or management pressure) to increase production. Such cases seem to be relatively rare in modern plants and in many cases are impossible because of trip settings. Cases from the days of steam power, of rail engine drivers and ship engineers tying down or gagging safety valves to achieve more pressure and speed, have been recorded in legends and in songs.

Errors in provision of amounts for batch processes appear relatively frequently in accident records. The most common is 'double charging', where an operator (or two different operators) add the necessary amount of a component twice. Other causes are errors in remembering appropriate amounts, errors in judging amounts and errors in measuring amounts.

ERRORS IN PROCEDURE EXECUTION—TOO LONG OR TOO SHORT DURATION

The duration of heating, cooling, mixing and grinding processes can be critical for batch processes.

A too short duration rarely appears as a cause in accident reports, presumably because endpoints are defined in terms of product properties, and it is usually possible to extend the processing period. By contrast, it is rarely possible to correct a product which is 'overcooked'.

Causes of too long duration are the following:

- Distraction, as described above.
- Hindrance, as described above.
- Forgetting. This is especially likely to be a problem when the operator has several tasks to perform.
- Misremembering, as described above. This is likely only when the task to be carried out is performed rarely.
- Mistake performing an operation for too long or too short a time because several different products are produced, with different production times. The actual product currently in production is misremembered or mistaken for another.

ERRORS IN ACTION PERFORMANCE

In actually executing the sequence of task steps, errors can be made:

- The procedure is not well known or not well remembered.
- There are two similar procedures, for example, for two similar units with slight differences.
- The labelling of the buttons, valves, etc., is ambiguous or unclear.
- There are simple lack of labelling and mistaking of two similar items.
- Two items are mistaken for each other because they are not arranged in logical order.
- The valve or button conventions or stereotypes are not followed.

- There are cultural differences in indication or actuation stereotypes, for example, right to left reading.
- The actuation, for example, the valve handle, is too stiff, is too slack or has a dead zone.
- The item to be controlled is too sensitive or has a dead time, making it difficult to control.
- There is lack of concentration, or mind wandering, thinking of something else.
- Feedback from the plant is inaccurate or missing.
- Known but poorly remembered hazard conditions are overlooked.

CASE HISTORY 5.8 Simple Omission of Check of Necessary Preconditions for Safety

An operator was to pump an active reactant (thionyl chloride) to a reactor, as 1 stage in a 20-step procedure. He did not remember to close other valves on the reactant manifold, so that the reactant flowed not only to the intended reactor but also to several others as well. The omission was almost certainly due to the fact that the operator had been recently transferred from another unit, where reactant supply was direct, not via a distribution manifold. The operator might never have learned his task properly, or he might not have remembered the changes needed when operating his new equipment.

CASE HISTORY 5.9 Operation on the Wrong Reactor [5]

Six operators per shift worked in the polyvinyl chloride 1 (PVC1) area where the incident occurred. Two operators, a poly operator and a blaster operator, were responsible for the reactor involved in this incident. The poly operator worked exclusively on the upper level of the building where the reactor controls and indicators were located, while the blaster operator worked on all levels.

To manufacture a batch of PVC, the poly operator readied the reactor, added the raw materials and heated the reactor. The poly operator monitored the reactor temperature and pressure until the batch was complete, then vented pressure from the reactor and told the blaster operator to transfer the batch to the stripper.

To transfer the batch to the stripper, on the lower level the blaster operator opened the transfer and reactor bottom valves. When the transfer was complete, the blaster operator closed the transfer valve, and the poly operator purged the reactor of hazardous gases to prepare the reactor for cleaning. The blaster operator cleaned the reactor by:

- Opening the reactor manway;
- Power washing the PVC residue from the reactor walls;
- Opening the reactor bottom valve (if not left open from the transfer) and drain valve to empty and
- Cleaning water to floor drains.

Once the blaster operator finished cleaning the reactor, he closed the reactor bottom and drain valves and gave the poly operator a completed checklist indicating that the reactor was ready for the next batch of PVC.

Because of the as-found condition of reactors, the Chemical Safety Board (CSB) concluded that the blaster operator probably went to the wrong reactor by mistake and tried to open the bottom valve to empty the reactor. The valve would not open, however, because the reactor was operating with the pressure interlock activated. Because he likely thought he was at the correct reactor, the blaster operator may have believed that the bottom valve on the reactor was not functioning. The CSB concluded that because the bottom valve actuator air hoses were found disconnected and the emergency air hose used to bypass the interlock was found connected, the blaster operator, who likely believed the reactor contained only cleaning water, used the emergency air hose to bypass the bottom valve pressure interlock and open the reactor bottom valve while the reactor was operating, releasing the contents.

The shift supervisor stated that the blaster operator did not request permission to bypass the interlock (required by procedure) or inform anyone that he had bypassed it. Consequently, operators attempting to control the release likely believed that a failure or other malfunction had occurred and tried to relieve the pressure inside the reactor to slow the release. The VCM (vinyl chloride monomer) vapor cloud ignited and exploded while the operators were working at the reactor controls.

The precise details of causality are unknown, because the blaster operator was killed. However, it does appear that after the initial mistake in identifying the correct reactor, even the strong indication from the unopenable valve did not cause the operator to rethink.

Risk-Reduction Measures for Errors in Executing Procedures

A number of methods for preventing errors in procedure execution are given in Chapter 21.

REFERENCES

1. J. Rasmussen, Accident Anatomy and Prevention, Risø National Laboratory, 1974.
2. U.S. Chemical Safety Board, *Investigation Report—Refinery Fire and Explosion and Fire*, BP Texas City, March 23, 2005.
3. A. B. de Souza Jr., E. L. La Rovere, S. B. Schaffel, J. B. Mariano, Contingency Planning for Oil Spill Accidents in Brazil.
4. M. Rogovin, *Three Mile Island: A Report to the Commissioners and to the Public*, Nuclear Regulatory Commission, Special Inquiry Group, 1980.
5. U.S. Chemical Safety Board, *Investigation Report: Vinyl Chloride Monomer Explosion*, www.csb.gov, 2004.

6 Operator Error in the Field

In a modern well-designed process plant, it is difficult to cause a major accident from the control room. Hazard and operability studies and safety integrity level reviews will have ensured that there is monitoring of all critical process parameters and alarms, trips and interlocks which will stop a plant or prevent hazardous events before they become serious. There are in principle possibilities to bypass trips and interlocks and to adjust trip set points beyond the safe levels. In a modern plant, these values are generally specified quite clearly, though, in the manual of permitted operations and in the standard operating procedures. In a well-designed modern distributed control system (DCS), overrides are possible only by means of a special key or password, and the fact of the override is reported automatically to plant management. Changes are subject to safety analysis through the management of change procedure.

In the field, by contrast, there is a very large scope for error, because operators in the field can open and close valves with no physical limitations (usually) and can start and stop motors, although these actions will generally be quickly reported via the DCS.

So what does a field operator do? A list of the typical actions is as follows:

- Monitors the plant for leaks, vibrations, clues to possible failures or abnormal conditions such as ice on valve seals, equipment left in inoperable state after maintenance, poor housekeeping, abnormal smells and noises
- Opens valves to allow start-up flows and opens and closes valves (valve lineup) to allow desired flows through manifolds
- Closes inlet and outlet valves for equipment which is to be shut down
- Opens vent valves for equipment depressurisation and opens drain valves for draining in preparation for maintenance or inspection; in general, isolating and deisolating equipment, although this may be done by maintenance teams
- Takes samples of process fluids or solids from sampling points or by using dip cups
- Starts, and especially restarts, pumps; pumps which are tripped because of power outages or faults on controls can often be restarted only from the field, in order to ensure that there is a degree of visual inspection before restart
- Couples up hoses to allow loading or unloading of tank trucks and control of liquid transfer (these actions are similar to those carried out from the control room for the process plant, but truck loading and unloading is only rarely carried out from the control room)
- Opens up valves and starts pumps to perform chemical injection (antifoaming agent, flux oil and corrosion inhibitor)

- Opens up valves and starts pumps for liquid transfer to batch reactors
- Adds solids from sacks and big bags, especially to batch reactors, and for some continuous processes
- In some cases, fine-tunes the opening of valves and the speed of pumps, to ensure correct process operation, i.e. performs the function of flow control

Each of these activities has characteristic error types.

OMISSION AND FORGETTING

CASE HISTORY 6.1 Omission of Action—First Example

An operator was supervising filling of tanks in a manually controlled tank farm. He started the operation, but then had to go to another area, almost 2 miles away. He became fully engrossed in the work there and did not realise the passage of time. Before he remembered the original activity, the tank had overflowed.

CASE HISTORY 6.2 Omission of Action—Second Example

A tank truck driver pulled up to a loading rack, coupled up hoses and started to fill the tanks with gasoline. Once started, he went to the safe smoking room to smoke (contrary to standing orders). He lost track of time. A tank overflowed, and the gasoline caught fire (from an unidentified ignition source), causing a major fire which destroyed the tanker and the loading rack.

CASE HISTORY 6.3 Omission of Action—Third Example

An operator was to drain down water from a waste oil–separation tank (the oil stays in the tank while unwanted water gradually separates out). The operation would take about 30 minutes, so once the drain-down was started, he went to the smoking room. There was less water in the tank than had been estimated, so oil began to run from the tank and passed to the separator ponds and from there to the community water-treatment plant. The water-treatment plant was shut down for several days, requiring wastewater and sewage to be transported by road tanker to another treatment plant.

CASE HISTORY 6.4 Omission of Action—Fourth Example

An operation supervisor was in charge of operations for an alkyd resin plant, with 15 batch reactors. After phthalic acid had been added to the reactor from sacks, fish oil was added, stirring and heating was started and the resin was left to 'cook' for about 15 hours. The supervisor was responsible for the timing, which required judgement, because the correct viscosity would be achieved only at a varying time to be judged by eye. The supervisor would often go home and return later, half an hour before the batch cooking time was complete. Over a 20-year period, several mistakes were made, which led to runaway and overflow of both resin and flammable solvent from the reactor.

The probability of this error is one of the lowest observed in process plant operations and indicates a very high level of involvement with the job. The fact that eight reactors were in use and generally had different start and finish times complicates the task.

ERRORS IN PROCEDURE EXECUTION—WRONG OBJECT OR MATERIAL

A characteristic form of error is performance of the correct action but on the wrong object. This is almost always restricted to cases where there are several items of the same type.

The primary cause of this kind of error lies in labelling or communication:

- No labelling
- Poor labelling with deterioration of labels, either falling off or fading
- Mislabelling
- Poor identification in instructions
- Error in identification of equipment in instructions

The problem is enhanced if objects such as pumps, distillation columns, etc., are not arranged in a logical fashion, with pumps A, C and B in a row, for example. Illogical layout is especially a problem for plants which have been extended, adding new units, new vessels or new pumps in the space available, leaving equipment out of order.

Operations on the wrong object can also be caused by errors in communication, particularly errors in written orders. Leading to error are not only writing the wrong name but also instructions which are not sufficiently explicit, e.g. 'start the cooling pump', rather than 'start pump 101-P-117A'.

CASE HISTORY 6.5 Wrong Ingredient [1]

One of the worst chemical accidents in history occurred as a result of a company running short of properly labelled sacks. The flame retardant polybrominated biphenyl (PBB) was packed into unlabelled brown paper sacks. The substance in powder form resembles magnesium oxide, which was also supplied as a feed additive for cattle. The elements of the accident were therefore readied and required only a mix-up in storage to trigger the accident. The PBB was supplied as additive to feed suppliers, who then supplied the feed to farmers.

PBB does not have an acute toxic effect on cattle in the amounts which were added to the feed. The effects are chronic. The poisoning was recognised as an epidemic some 6 months after the original distribution of the contaminated feed. Cattle began to fall, to crawl on their knees and to die. At first, the cause of the 'disease' could not be identified. A chemist from a Michigan public health laboratory tried to determine possible contamination of feed by using liquid chromatography (note that at this time it was not even known whether poisoning was the cause). An unknown substance peak was seen in the chromatograph output. The chemist had coincidentally analysed residues a few months earlier and so was able to identify the substance.

Altogether about 30,000 cattle and 1.5 million chickens were poisoned, and an unknown number of persons ate contaminated meat and drank contaminated milk, with human doses of up to 10 grams.

Labelling problems or lack of knowledge or a combination of both can lead to errors in use of the wrong substance.

A problem with fine chemicals or speciality chemicals is that a given chemical can have many different names. These can include variations in the formal chemical name and differences in trade names. To complicate matters further, production companies often abbreviate the names of the chemicals by using acronyms. *IPA* can, for example, stand for *isopropylamine*, *isopropionic acid* or *India Pale Ale*!

Even when chemicals are properly labelled, mistakes can be made by operation assistants or plant labourers with limited chemical knowledge. It was once asked, for example, if there was a difference between potassium chlorate and potassium chloride. Given the variations in trade names and lack of an education in chemistry, this is not an unreasonable question. It may surprise some engineers and plant managers, but in some companies, raw materials are in some cases recognised by packaging type and colour, something which is almost guaranteed to cause problems at some stage, when packaging changes or when the wrong product is delivered. A good procedure for receiving materials is needed, with a purchasing and a warehouse manager with at least a degree of training, for any chemical plant. Also, all personnel handling chemicals should be taught labelling practices and substance recognition, as well as the hazards of spills and safe cleanup techniques.

CASE HISTORY 6.6 Wrong Equipment Material [2]

Chlorine was unloaded from railcars by using alloy-reinforced 1-inch hoses. Because chlorine corrodes most steels, including 316 L stainless steel, the reinforcement required is Hastelloy, an alloy of nickel.

In the actual event, hoses reinforced with stainless steel were supplied. The first one used corroded within a few tens of minutes. There was no emergency shutdown valve fitted at the railcar end of the hose, and the excess flow valves fitted were, as is normally the case, not sufficiently sensitive to close in the case of a hose rupture. (Railcar excess flow valves are usually designed to close in the case of valve damage in a rail crash.) As a result, in all 22 tonnes of chlorine were released.

CASE HISTORY 6.7 Wrong Hose Connection

A frequent location for errors of object choice is in tank truck unloading onto storage tanks. When fluids are pumped into the wrong tanks, a reaction can occur immediately or later when the liquid is used. Examples which are known in accident records are the following:

- Nitric acid (used for piping sterilisation in biochemical production and brewing) pumped into formaldehyde (also used for sterilisation), causing an explosion
- Hydrochloric acid pumped into sodium hypochlorite and vice versa, causing a release of chlorine and in some cases tank rupture
- Low–flash point solvent or gasoline pumped into fuel oil tanks, with a potential for explosion in the tank or at boiler where the fuel oil is used

The logically best solution to this problem, of using different size hose connectors for different substances, has proven to be impossible to organise except in a few isolated cases where deliveries are routine. Even with this solution, accidents have happened in which truck drivers have fitted adaptors to hoses to allow connection to the wrong-size receiving nozzle.

Contributing to this kind of accident have been poor labelling of tanker unloading nozzles, lack of an operator to supervise unloading and lack of written instruction for truck drivers. Well-designed unloading stations have locks on unloading nozzles, so that it is impossible for a truck driver to unload without supervision.

Case History 6.8 records an incident in which a field operator shut down one of two pumps.

CASE HISTORY 6.8 Wrong Pump

A power station boiler had two high-pressure feed water pumps. In order to avoid the need for high-pressure seals, the pumps were provided with pressurised motors. The windings, although enclosed, were subject to the high pressure of the boiler feed water.

It is one thing to pressurise a pump, and another thing to overheat it. To avoid the effects of very hot water on the motor, the design required high-pressure cooling water to be supplied between the pump and the motor.

An operator was told to shut off the flow from a seal water drain line on one of the pumps (Figure 6.1). He went to the pump, and shut off the cooling water flow. With the pressurisation gone, hot boiler feed water passed backwards through the pump seal to the motor and destroyed it. The result was a shutdown and a loss of several million dollars for the new motor and for the loss in production. The operator had drained from the wrong pump, the one that was not shut down.

When the pump arrangement was looked at, the reason for the mistake was not too difficult to see. There were 25 pipes and hoses intertwined and arranged

FIGURE 6.1 Two canned high-pressure pumps. Cooling water to the wrong one was shut off, resulting in pump destruction.

according to the dictates of efficient flow, not operating logic. The first impression was one of steel spaghetti. The valves, though, were properly labelled. Labels were moderately difficult to read, not being placed at eye level. Lighting was neither good nor bad; it used electrical lights in a pipe tunnel.

The opportunities for mistake were obvious—several valves were side by side, allowing confusion. The operator had performed similar tasks earlier. Possible direct causes are misremembering of location and confusion of the actual identity of the valve. Contributing factors could be reading difficulties. After the event, the operator had difficulty explaining the exact cause.

A normal risk assessment would identify the opportunity for opening the wrong valve and identify poor labelling, inadequate lighting, proximity of similar but unlabelled valves, etc., as performance-influencing factors. The action error–analysis method in Chapter 19 would identify proximity, misremembering, confusion, lack of knowledge, label misreading and overlooking the label as direct causes and poor lighting as a performance-influencing factor.

There is one hidden contributing cause to this accident. That is, the operator almost certainly did not realise the possibility of a critical mistake and the degree of consequences. As a result, he almost certainly did not apply the degree of care in cross-checking, which should have been required.

In order to provide the operator with the proper background, today the following would be regarded as good practice:

1. There was a proper hazard analysis which identified the problem of loss of cooling.
2. The hazard was identified in the operation procedures as a warning.
3. The hazard-analysis results were incorporated into operator training.

In other similar situations, operators have been afraid to ask for confirmation or support when unsure or confused, because they did not wish to reveal their lack of experience. The social climate in a work team can affect error rates significantly. Unwillingness to seek advice or confirmation can stem from teasing or mild forms of bullying; feelings of inadequacy on the part of the operator and tension arising from the employment situation, especially when downsizing is being considered.

Establishment of awareness and a good team spirit of mutual support are questions of leadership and people management. They can be taken into account in auditing and can even be factored into risk assessment, as shown in the next section.

TOO MUCH FORCE

Excessive use of force is often associated with trying to close valves to prevent passing, trying to open stuck valves or trying to tighten flange bolts to stop leaking.

CASE HISTORY 6.9 Excessive Force

Operators were trying to tighten flange bolts on a high-pressure water pipe, to stop a leaking jet of water from a valve flange (very dangerous; the pressure was so high that the water jet could kill). A ring spanner (ring wrench) was used, and then an extender, i.e. a length of scaffold pipe, was added to give more torque. The torque was so high that the valve broke. A jet of water blasted from the valve, hitting an operator and throwing him across the platform. He was killed when his head hit the platform railing.

This kind of macho get-things-done behaviour is common in many plants. The operators were trying to solve a problem without shutting down the water injection to an oil field. In fact, a shutdown and pressure relief in this kind of operation does not have a big penalty, and the job could well have been done more safely.

Too much force is a common cause of minor to severe injuries during maintenance. The 'extended wrench' consisting of an ordinary ring wrench with an extended handle made from a 1/2 to 1 metre length of scaffold pipe is a really useful tool. It will remove flange nuts which have rusted and open valves which have 'frozen' due to corrosion or deposits. It can also, in enthusiastic hands, shear bolts, crack valve operating wheels from their spindles and rupture steel piping up to about 2 1/2 inches in diameter.

REFERENCES

1. G. F. Fries, G. S. Marrow, and R. M. Cook, Distribution and Kinetics of PBB Residues in Cattle, *Environmental Health Perspectives*, Vol. 23, pp. 43–50, Apr. 1978.
2. U.S. Chemical Safety Board, *Investigation Report, Chlorine Release*, http://www.csb.gov, 2003.

7 Knowledge and Ignorance in Plant Operation

As noted in Chapter 5, there is an enormous difference between operating a plant by following procedures and operating a plant when you understand and know what is happening. In fact, most experienced operators have a fairly good knowledge of how a plant works and track what is happening in the plant by 'mental simulation' and by checking the correspondence between feedback and expectation. In many cases tracking is done instinctively, rather than consciously.

CASE HISTORY 7.1 Instinctive Recognition versus Knowledge-Based Explanation

An example of this was observed in a control room for a small distillation plant. An experienced operator suddenly got up from the operation panel and went outside to observe a level gauge and a motor. He had expected a product tank to fill and a transfer pump to start. This had not happened as expected, and on investigation it was found that the product/reflux valve was stuck in the full reflux position; i.e. there was no production. All the distilled product was being cooled and returned to the distillation column.

As this example shows, the operator knew the process and either was integrating the production rate subconsciously or was running some kind of 'internal timer'.

The knowledge of the process was not complete, however; the disturbance due to the reflux valve sticking was fairly obvious from the column head temperature, for anyone knowing the distillation process, and could have been recognised as the problem developed. The degree of skill in recognising the plant disturbances is dependent on both the degree of knowledge and the ability to use that knowledge. The operator might have learnt to associate the event 'no product' with full reflux, or might be able to work out the cause by thinking about how the column works (see 'Backwards Simulation' section in Chapter 10).

(Note that different columns will respond to disturbances in different ways, depending especially on the control scheme used, and that column response needs to be learned, even for experienced operating engineers.)

There can obviously be several degrees of knowledge, and the more knowledge available, the more that the operator can predict or check. There are also different types of knowledge. The following types have been found useful in accident investigations and predictions:

- Knowledge of basic physics
- Knowledge of substance properties
- Knowledge of the actual plant construction, layout and working
- Knowledge of the actual state of the plant
- Knowledge of disturbance and accident physics

When something occurs within a plant which is not covered by procedures, knowledge is needed to determine what is happening and then to decide what to do about it. An example of this gives some idea of the range of knowledge involved.

CASE HISTORY 7.2 An Operating Mystery

During commissioning of a batch distillation plant, the team, of which the present author was a part, could not get the temperatures right in the overhead condenser. The condenser was a shell and tube heat exchange and was required to have a low temperature at the start of distillation, when methanol would distil off, but to be hotter at a later stage, when water would come off. Both hot water and cold water could be supplied to the condenser in order to avoid return of a too cold reflux stream to the distillation kettle during the hot phase of distillation. Three-way valves were provided to allow this.

The difficulty in achieving the correct temperatures was a mystery which puzzled us for a few hours, but there was obviously something wrong with the flows. A complaint was received from neighbouring plants' operators that our plant was taking all the hot water, but we could not get a high enough temperature in the condenser.

I checked the installation of the three-way valves and found that one of them was installed the wrong way round. Either cold water was being supplied or hot water was being discharged to the cooling water supply. The valves looked to be properly installed, side by side, but the direction arrows on the valve bodies were pointing the same way and needed to be opposite. What was obviously wrong when expressed as a piping drawing was far from obvious in the plant itself.

The origin of the knowledge required for identifying this problem (and getting it corrected very quickly, just by installing one valve) was very clear. My first job in a precommissioning troubleshooting team of a nuclear power station was to check the direction of flow in all the valves. The idea that a valve could be installed the wrong way round came very readily.

Based on the idea, it became necessary to be able to trace what a wrongly installed valve could do. This was done using a sketch of the piping and then a mental simulation of how it would work with different valve positions.

These types of thinking are very standardised and stylised for experienced operators and for process engineers. The types of knowledge commonly used have been identified in many postincident and troubleshooting reviews:

- Knowledge of 'problem prototypes', such as reverse-mounted valves or pump cavitation.
- Knowledge of physical phenomena, such as water hammer.
- Knowledge of how components behave, such as fall in pressure at the discharge of a centrifugal pump when the flow is increased.
- Knowledge of how disturbances or changes propagate through a process system. A simple case is what happens when flow increases into a vessel. Usually the level rises and then stabilises as a level controller kicks in. A more complex example is what happens in a distillation column when the feed preheating system fails because the steam valve is closed. (This kind of disturbance tracing will be discussed more extensively below.)
- Knowledge of how the actual plant is piped up and what the equipments are.
- Knowledge of the actual state of the plant at the time when action is required.

In many follow-up studies of hazard and operability (HAZOP) analyses, the observations have been that in order for problem prototypes and disturbance phenomena to be remembered, (1) they must be given a memorable and unique name and (2) the name must be associated with a clear picture.

An example of this is the occurrence of boilover and fire-induced tank explosion, which are often confused in the literature, even though they are completely different phenomena. The confusion occurs because the second term is not a recognised one, so the first is used ambiguously. The emergency actions are correspondingly wrong.

KNOWLEDGE-BASED OPERATION UNDER MODERN CONDITIONS

Control system designers today try to eliminate the need for knowledge-based operation. Alarms, trips and interlocks are provided to ensure that a system cannot get into a dangerous state. Many designers believe that such systems will be completely effective. In HAZOP studies, the belief that 'accidents cannot happen because we have provided a safety valve/high-level trip/check valve, etc.' is often proved wrong. These days it usually does not take too long to persuade HAZOP review participants to say, 'The probability of that happening here is very low because …', but the implication is still there, that designers do not really believe that their designs can fail and definitely not that these can kill. Perhaps if designers did, they would not be able to design well. Nevertheless, accidents continue to happen, sometimes for quite complicated reasons.

The protection provided by modern safety systems has several impacts on operators. One is that they have much less opportunity to learn how to deal with disturbances—there are fewer of these, and if they do occur, they will often be dealt with automatically.

A second effect is that the operator is today much more isolated from the plant by a layer of controls which filters out a lot of detail. The operator 'sees' the plant through the supervisory control and data acquisition or distributed control system.

The mode of thought of an operator at a workstation can be very different from that of an operator who works in the plant.

An experienced operator working in the field can see through steel. The sixth sense which allows this is composed of intimate knowledge of the equipment, together with the process of sensing of the noises, vibrations and bad temperature of the equipment. For example, if a propane storage vessel shows a trace of frost on the shell, this means that the safety valve is leaking internally.

Interviews with operators who can do this show that they actually have a 'picture' in their minds of what is happening inside the equipment. Further questioning showed that this picture was largely related to just one item of equipment at a time. However, the operators could follow the flow along pipes into new vessels and through equipment, visualizing along the way.

Interviews with process engineers show that many have the same ability, especially if they have some experience of plant operations. However, the visualization is not specific to an actual plant. For example, an operation supervisor walked into a pump cellar and immediately walked to a valve and operated it. Then she continued by lecturing an operation assistant about remembering to open pump suction valves before starting a pump. She had been able to hear the rather faint sound of a screw pump sucking air through the pump seal. Few design engineers would be so sensitive to faint sounds or the specific properties of that actual pump.

Interviews with panel operators/workstation operators showed that some do think in this way, but many more think in terms of expected parameters, deviations of parameters and possibly balance of flows.

KNOWLEDGE OF PLANT STATE

Knowledge of the actual state of the plant can be critical for safety, even when very experienced operators are in control. An example is the accident at the Piper Alpha Platform in 1988, where an operator repressurised a system, even though one of the piping flanges had been left open. He simply did not know what kind of plant he was running; it had changed since the previous day.

AMBIGUITY OF INPUT

Sometimes indications from the plant are ambiguous. At some point this will always be true. A high pressure in a distillation columns feed tank, for example, can result from an outlet valve blockage or closure, an inlet valve failure or a high feed temperature. In a highly instrumented plant, there will be sufficient instrumentation to disambiguate the problem. With less instrumentation, the board operator may need to ask a field operator to check valve positions.

Even if there are sufficient inputs to clearly indicate a problem, it will not necessarily be the case that the operator makes the right conclusion. Operators often control the plant on the basis of a few key parameters, such as distillation column pressure drop, column overhead temperature and separator level. The emphasis is in this case on securing good performance when disturbances occur; the operator should look wider but often will not.

KNOWLEDGE, BELIEF AND JUDGEMENT

Beliefs are established by learning, either from others or by experience. Some of these beliefs can be hazardous or even immediately lethal if the correct circumstances arise.

CASE HISTORY 7.3 Deliberate Shortcutting Due to Lack of Awareness

A young technician was asked to install a light fixture. Clear instructions were given to open the main group circuit breaker. He judged, however, that opening the actual lighting circuit switch was sufficient, allowing other lights to be used for working. An older technician insisted that the younger technician test the voltage, after sparks had been observed. In fact, the circuitry was miswired, with the light switch in the neutral (current return) line, not in the power line and the colour coding was completely misused (actual power was supplied via a yellow wire).

Erroneous beliefs, as opposed to lack of knowledge, can lead to accidents. Two fitters were to replace a seal on a pump. They closed the upstream and downstream valves, believing this to be adequate. They then opened a drain valve. Hot oil was released and caused a fire. The reason for the remaining pressure in the pump was a bypass line from the pump discharge to the feed tank, which had a partially failed pressure relief valve.

Case History 7.3 concerned belief about the status of wiring.

Other examples of hazardous beliefs are the following:

- Potential rescuers often believe that they can rescue colleagues who are overcome by poisonous gas just by holding their breath. This often leads to multiple fatalities. Hydrogen sulphide fatalities, for example, often come in threes.
- Some operators believe that there is always a potential overpressure safety margin, that they can operate at higher filling or throughput rates (safety factors are in fact used to account for inaccuracies in theory, manufacture, installation effects, etc., and not to provide extra capacity to help meet production quotas).
- Low-pressure steam is often regarded as not dangerous.
- Concrete is often believed to be a nonhazardous substance. A construction worker waded in concrete, which overtopped his boots. He completed the distribution of the pour before removing his boots. Feet and legs suffered alkali burns.

A particular group of hazardous beliefs is that a plant is in a normal or safe state. One of the main differences between a cautious and an unlucky operator is the tendency to check rather than believe.

CASE HISTORY 7.4 Wrong Understanding of Plant State

An engineer was supervising a series of tests which involved dosing liquids to an incinerator by using diaphragm pumps and hoses. During setup, tests were made with water. Several hose bursts occurred. During a later run, the engineer asked whether a particular hose had been tested. The maintenance fitter confirmed that everything had been tested and that there was only water in the system. When the pump was started, the hose was forced off a pipe nozzle, and the engineer was covered in toxic solvent.

The fitter believed that an appropriate test had been made and that the tank contained water. In fact, the test was made with an open discharge valve, not a closed one. Also, the feed tank did contain water but only on top of a small amount of heavy solvent. The fitter believed that all solvents are lighter than water. The solvent entered the pump first.

In everyday work, operators must believe in some things. It is necessary to believe that pumps will work, tanks will hold liquid and so on. Otherwise, operation becomes impossible. It is not feasible to check everything at every instant in time. Appropriate caution involves knowing which beliefs and assumptions have a probability of being wrong and which are critical for safety.

Caution of this type may be taught as cautions and warnings in procedures. Cautions and warnings are not always as effective as procedure writers hope, however. Operators do not always read them or understand them, and they do not always remember or believe them. Engineers, in writing procedures, are generally very poor at explaining why something must be done or not done. Standard operating procedure formats do not require or encourage this. This more or less invalidates written operating procedures as teaching aids—in tests our teams have found that retention of knowledge improves by between 5 and 12 times if explanations of reasons are given and is further enhanced if the phenomenon which results from doing something wrong is given a clear name and is accompanied by a picture.

In particular it is often difficult to inculcate the deep-seated belief which is necessary for consistently safe behaviour. An example is the cautions against leaving a tank truck during filling. Many such trucks have suffered overflow when the driver left the truck to fill to go away to smoke a cigarette. Several beliefs can contribute to accidents here:

- The belief that the driver can judge the filling time
- The belief that the high level shutoff will work
- The failure to believe that unsupervised filling is dangerous

The most direct and effective source of beliefs is experience. Once an operator has been involved in a fire, for example, the quality of belief in solvent or oil hazard changes considerably. It is possible to change beliefs quite effectively, by teaching based on accident case stories and using photographs.

An experienced operator or maintenance technician will make decisions based not only on theoretical performance of equipment but also on what can go wrong and what can be wrong. Caution of this kind in operations is often learned through painful experience.

CASE HISTORY 7.5 Erroneous Belief in a Perfect Plant

As a further example, an experienced maintenance fitter was required to replace a valve in a thin (1/4 inch) chlorine pipe. The procedure involved the following:

1. Shutting down the system
2. Waiting 2 hours for all chlorine to evaporate
3. Closing manual shut-off valves at either end of the 1/4 inch line
4. Cracking open the flange on the control (loosening bolts and) valve
5. Waiting for any chlorine escape
6. When no chlorine was seen to escape, replacing the valve

The fitter was impatient and decided just to close the manual valves and to release the chlorine at the flange. Unfortunately, one of the manual valves had been specified to be bypassed by a safety valve, to provide pressure relief. Still more unfortunately, the actual equipment installed was a graphite burst disk, and this was cracked.

When the fitter opened the flange, a considerable quantity (80 kilograms) of chlorine escaped.

Counterexperience can be quite dangerous in destroying true beliefs. Very often during audits and hazard analyses one hears the statement: 'That cannot happen; we have actual experience'. An example is the possibility of heat exchanger rupture due to leakage in which a light liquid fraction mixes rapidly with a hot, heavy fraction and vaporises. Small leaks seldom lead to rupture of the heat exchanger, but small leaks are by far the most frequent type. Operators learn by experience that this is only a cause of operational disturbance. In fact such leaks can lead to catastrophe, with the heat exchanger or the piping rupturing. As an extreme example, such a leak occurred, forcing liquid into the heat exchanger discharge pipe and causing an overload on the pipe which ruptured and fell across the terminals of a live 10 kilovolts transformer, with predictable results.

Safe operation in plant often depends on knowledge, beliefs and judgements. As in the examples, knowledge is needed, not only about the system and what can be wrong but also about what can go wrong. The amount of knowledge needed is often quite large. The following is a list of both common and arcane facts which operators need for safe plant operations:

General knowledge which can save lives

- High flow rates can cause erosion.
- Starting a pump can cause hammer.
- Opening a valve to an empty line can cause hammer and rupture. Piping must be filled gradually.

- Closing a valve can cause hammer. Vessels, heat exchangers and low points in piping can contain liquid which is not expected. When steam flow starts, liquid slugs can cause hammer. Hammer can rupture pipes.
- Filling a pipe with liquefied gas can lead to bubble collapse when the pipe fills, and this can cause hammer.
- Vertical two-phase flow can cause heavy vibration.
- High flow rates can cause vibration. Vibration can cause fatigue cracking.
- High flow velocities, bends and expansion can cause cavitation and erosion in pipes, not just in pumps.
- Liquids can foam. Foam flow gives rise to high pressure in vapour or vent lines.
- Blockages in pipes can give rise to high pressures. Removing a blockage can be dangerous.
- Pipes can burst if high-pressure air or steam is used for cleaning.
- Plastic pipes can burst if any air pressure is used to clear blockages.
- Residue in filters can be toxic.
- Filters can catch fire due to impurities.
- Filter pots can be pressurised and pressure gauges and bleed valves can be blocked. When opened, the lid can fly off. The flying lid can take off your head.
- If the vent on a tank is blocked, pumping liquid from the tank can cause collapse. If the tank is more than a couple of metres high or the exit line falls by several metres, just the static head can be sufficient to cause a vacuum.

A LITTLE KNOWLEDGE CAN BE DANGEROUS

CASE HISTORY 7.6 A Little Knowledge

After a leak of acid in a soft drink factory, a fire brigade commander, knowing that alkali would neutralise acid, told a fireman to add caustic soda. The fireman was killed when the mixture boiled violently due to the heat released from mixing acid and caustic soda.

An obvious and direct cause of error is erroneous or inadequate knowledge, or lack of knowledge. A characteristic of situations in which knowledge is necessary is that they are unusual—that either they require improvisation or the operator deviates deliberately from procedure.

PLANNING NEW PROCEDURES

Situations can arise for which there are no procedures. (Clearing blockages, for example, is a difficult task with many variants, often requiring experimentation.) In this case, operators and maintenance technicians are forced to develop new procedures.

Deviations from procedures can arise because the operator wishes to save time or effort. Deviations can also arise because the procedure is inappropriate or just plain wrong. If the written procedure is wrong, the operator is forced to find a different way of doing the job. The revised procedure will rarely be checked for hazard by means of HAZOP analysis, as it should be prior to plant commissioning.

When procedures are changed, the changes should be marked up on a master copy and should be subject to change control. In modern practice, before this is done, any new procedure or procedural change should be subject to job safety analysis or HAZOP analysis.

THE NEED FOR JUDGEMENTS

Judgement is often necessary in plant operation, particularly in unusual operation conditions.

An example of the need for dynamic judgement occurs in ship navigation and concerns incipient collision between two ships. The navigator needs to judge the paths of the ship and the other ship and judge whether reversing engines, sailing in front of or behind the oncoming or crossing ship, or sailing in a loop will be the safest. At the same time, it is necessary to judge the likely actions and performance of the oncoming ship (although further information can be gained by radio).

An example of dynamic judgement in process plant occurs with a chemical reactor which is overheating. Will cooling be adequate to regain control, or will it require quenching or dumping the batch? The projected heat-up curve needs to be judged. The operators often have experience only on which to place the judgements, and this experience will in virtually all cases be incomplete, because it cannot cover all possible circumstances. Designers have, in contrast, access to underlying theory, calculation models and all the standards governing the design but only rarely experience from 'real life'.

Dynamic judgement requires experience, learning the rate at which changes take place, its accelerations and decelerations and an ability to integrate these. Judgement of this kind is similar to that for manual control but with the opportunity for training and learning of specific situations much reduced. (It is also the same kind of skill which is needed for playing many computer games, but without the possibility of pressing the restart button.)

Parametric judgement is involved when someone needs to estimate the value of some parameter which cannot be measured directly. Typical of such judgements are the following:

- How much liquid is left in a tank, given that the level gauge has failed
- How large is the hazard area around a vapour leak
- Whether a lifting beam will take the necessary load without failing

Such parametric judgement can sometimes be based on calculation if the necessary basic data are known. Such judgements must often be made quickly, however, or when there is insufficient data, and estimates must be relied on.

CASE HISTORY 7.7 Erroneous Judgement Due to Too Little Knowledge

In the restart of the bromination reactor described in Chapter 15, operators needed to estimate the speed of reaction if stirring were to be started in an over-dosed reactor. Unfortunately they underestimated the rate, probably underestimating the importance of 'self-stirring' due to local boiling.

Probabilistic judgement is needed when there is uncertainty about performance. For example, will a valve which has not been used for some years still work? Or will an operator be able to control a particular flow accurately enough to prevent a trip?

Judgement depends on both knowledge and belief. But judgement additionally is sensitive to judgement skill and to priorities. It is often observed that if the operations manager sets a high priority on production, judgements will be optimistic. A more cautious manager will make more pessimistic judgements. These judgmental effects appear before the stage at which risks are balanced and decisions made.

The need for judgement is not only restricted to operators. Plant design engineers, quite often, are also faced with unusual situations.

During HAZOP studies, it is often necessary to make judgements, particularly about the physical possibility of events and about their extent. A sufficiently all-knowing engineer with ready ability to calculate and simulate effects in principle could avoid the need for judgement. In practice, even with this ability, there is often insufficient time or motivation to make detailed calculations.

CASE HISTORY 7.8 Lack of Knowledge in Design—Hazardous Property of Materials

A team of engineers had carried out a HAZOP study of a hydrochloric acid burner and recognised the possibility for burn back in a chlorine/hydrogen mixture but were unaware of the possibility of detonative explosion. The piping was not dimensioned for detonation pressure.

CASE HISTORY 7.9 Lack of Knowledge in Design—Disturbance Condition Not Known

A team of engineers dimensioned a knockout drum for a vent system. They overlooked and initially found incomprehensible the fact that the system would fill with foaming resin, and the knockout drum would require 20 times the volume and 4 times the pipe diameter to be able to vent the foam (a case for 'back to the drawing board').

CASE HISTORY 7.10 Lack of Knowledge in Design—Failure to Believe That a Design Simply Cannot Work

A similar problem which can arise is in determining the size of burst disks for a chemical reactor. The theoretical calculations often lead to sizes which are larger than the reactor itself. Engineers find this difficult to believe, having seen many burst disks of reasonable size. The problem in these cases is that the reaction can be too fast, and the burst disk is an unsuitable approach to safety. The reaction which cannot be vented must be prevented.

CASE HISTORY 7.11 Lack of Knowledge in Design—A Mystery Which Remains Unresolved

As a fourth example, an explosion occurred in a gas oil tank because of carry-over of hydrogen from a stripper unit. Even after detailed investigation, it proved difficult to believe that hydrogen could dissolve sufficiently in gas oil and then could come out of solution in storage. And it proved equally difficult to believe that hydrogen could accumulate in a top vented tank. Only after confirmation by searching the literature and finding a record of a similar explosion in another plant could all engineers be convinced. If such problems arise after the event, how much more difficult are judgements made during predictive studies such as a HAZOP study.

One way of overcoming such otherwise unavoidable ignorance is to study accident records. However, unless these are both systematised and generalised, so that each physical effect is noted, the record of accidents will be too limited to be a reliable predictor of the full range of consequences. As an example, a large database of 10,000 accidents in chemical plants was searched and yielded only four examples of an explosion in a distillation column.

If accident records are studied and physical phenomena are isolated, these can be studied and generalised. Ref. [1] gives descriptions of some 250 accident mechanisms which operators may need to know about in order to be able to operate safely. In any one plant, only a fraction of these may be relevant, but the fraction may be a large one.

CASE HISTORY 7.12 (Lack of) Knowledge and Judgement

As a further example of erroneous operations, consider the following example. An operator had to feed liquid from a drum to an incinerator. Appropriate piping valves and hoses were available, but the pump designed for the application had failed and was under repair. The supervisor, therefore, told him to use another pump, which, unfortunately, was overdimensioned. The operator was told to be very careful in adjusting the discharge valve, to avoid too rapid feed.

When operation began, the pump being overdimensioned and nearly blocked in (nearly deadheaded) by the discharge valve burst the hose. After two burst hoses, the operator realised that the problem was that there was too much back pressure and opened the discharge valve fully. The feed rate became very high, causing a furnace explosion.

This example shows several aspects of human error. One is the problem of performing normal operations under abnormal conditions. A second is the error of modifying a plant in a seemingly harmless way. Note that operations supervisors need to be able to carry out this kind of adaptation regularly and need to be able to judge the degree of change involved.

KNOWLEDGE OF PLANT PHYSICS

What you do not know can kill you. Among the things you need to know about to operate a plant safety is a great deal about practical physics.

CASE HISTORY 7.13 Lack of Fundamental Knowledge of Physics

During a strike at a coke plant, the management staff were required to operate the plant. (This was in the 1970s; few managers would be capable of operating today's complex plants unless they have operating experience themselves on the actual plant.) A junior manager noticed some coal wagons running away slowly down an incline. He tried to stop them by pushing. The wagons pushed him slowly backwards until he was crushed against a buffer. The manager had not understood the difference between speed and momentum at the instinctive level, which would have led an experienced operator not to even try pushing a coal wagon.

In the next section some of the physical effects which are important in plant operations are described, including examples of accidents to which they have led. Many of these effects are common knowledge for design engineers but may not be known to operators. Some of them are rarely known even to design engineers (see the example of liquefied gas hammer below). The effects may happen only rarely, but when they do occur, they can kill. If anything, operators are more in need of this knowledge than the design engineers are, because they are the ones needing to deal with abnormal situations.

Hazard awareness courses help to some extent in providing safety information to operators, but hazard awareness courses are not yet commonplace in all companies. Also, currently available hazard awareness courses (in 2012) are in virtually every case extremely limited in scope. A survey of 40 process industry commercial hazard awareness courses showed that some focussed on safety regulation; some focussed on hazards in the workplace, such as tripping hazards and fire hazards and some focussed on accident consequences such as jet fires, boiling liquid expanding vapour explosion (BLEVE) and vapour cloud explosion. The only courses found which focussed on the process, and the operator's control over the process, were those on runaway chemical reactions. During the survey a few excellent web-based company-specific hazard awareness courses were found which dealt with some site-specific hazards.

An extended list of accident phenomena is given in Ref. [1].

AN EXAMPLE OF ACCIDENT PHYSICS—MOMENTUM

Moving objects, solids or liquids and even gases, possess momentum. When the movement is stopped or changes direction, considerable forces develop. A car moving at 80 kilometre per second has a momentum of about 112 tonnes · metre per second. If it hits a concrete bridge pillar, it will be destroyed. A small ship moving at 1 metre per second towards a dock has about 20 times the momentum of the car and will probably destroy a good deal of the dock if it hits square on. The oil in a 50 kilometres long 10-inch pipeline moving at 3 metre per second has about 5000 tonnes · metre per second of momentum and will smash most equipment if the flow is stopped suddenly. The effect is called hammer, and there are several different kinds about which the operator will need to know, depending on the plant type:

1. Pressure rises rapidly in a pipeline containing flowing liquid if a valve is closed quickly. The overpressure depends on the speed of closure, the velocity and the length of the pipeline. The pressure developed can destroy the pipeline or the valve if the flow is fast enough and the pipeline long enough. A pressure wave can also pass backwards up the pipe and damage equipment such as instruments. For this reason valves in long pipelines are designed for slow closure.

   In a large and long pipeline, safety valves were provided at pumping stations, so that if emergency valves closed, the hammer pressure would be limited by redirecting the flow for a short period to a surge tank. The spring supports and dampers on the surge line were not properly adjusted, so the full force of the jump of the surge line was put onto the tank entry nozzle. The pipe tore away from the tank and released a large amount of oil into the bunded area.

   Note that responsibility for proper adjustment of pipe support springs, dampers and snubbers is the responsibility of commissioning engineers, but operators or supervisors and integrity teams will need to inspect these in their mechanical integrity inspections and maintenance technicians would need to be able to adjust these properly. In the case described above, the springs had originally been adjusted, but only prior to operation, so that the supports were not adjusted for the case of an oil-filled line.

   This kind of hammer will not occur in a gas pipeline, because the gas is readily compressible and simply reduces in volume as the pressure rises giving a cushioning effect.

   Operators need to be aware of this kind of hammer effects, especially when the pumps feeding a pipeline have sufficient capacity to allow operation above the design value.

2. If a pump feeding a previously empty pipeline is started with an open discharge valve, the liquid will flow rapidly against little resistance and can reach a high velocity. If the flow reaches an elbow or a restriction, a high force will develop, which can be sufficient to rupture the line.

CASE HISTORY 7.14 Lack of Awareness of Hazard [2]

A multiproduct pipeline was being refilled with kerosene after cleaning and flushing with nitrogen. The pumping was carried out at full pump capacity, since initially there was no back pressure. Unfortunately the battery limit valve at the receiving terminal had been left almost closed. When the kerosene reached the valve, it shattered due to the hammer effect. The operators kept the pump in operation for over half an hour until the kerosene pool was noticed by the operators at the receiving terminal. (In modern practice there should be a level transmitter which indicates the receiving tank level to the operators controlling the pump, and there should be a trend display. Also, for any critical line there should be at least a primitive mass balance leak indicator.)

Ironically the rupture would not have occurred if the valve had been left fully shut (because of the gas cushion).

A similar kind of hammer effect can occur if a pipeline with a closed discharge valve is being filled and a vent valve is used to release gas in the pipeline. When the line fills, the liquid must flow through a narrow vent line, and hammer damage can occur if the vent line, which usually has a small diameter, cannot take the liquid flow. For this reason, the amount of filling should be tracked and the rate of filling reduced when the pipeline is close to becoming full.

3. If a pump trips, or a pump discharge valve fails closed, the momentum of the flow can cause a vacuum. This is especially important for large-diameter water lines such as cooling water headers, where the vacuum can cause pipeline collapse.

   Operators and maintenance technicians need to know this effect and be observant for proper operation of vacuum valves when pumps are shut down or discharge valves close.

4. Two-phase flow lines, gas lines which have condensate drop out, for example, in cold weather, and steam lines in which water condenses can develop stratified flow. A layer of liquid flows or collects at the lower part of the pipe, and the gas flows above. Often, the gas flows much faster than the liquid.

   This kind of flow can be stable until the liquid level rises too high or the gas flow rate increases too much (or both, because collecting liquid partially blocks the pipe and leaves less area for the gas to flow). When either the liquid level rises too high or the gas flow velocity increases too much, waves form on the liquid surface. If the wave height is sufficient to reach the top of the pipe, gas flow is blocked by the liquid, and the gas then pushes all

FIGURE 7.1 A pipeline affected by condensate hammer.

of the liquid in front of it. This is slugging flow. This can be quite dramatic, because the gas flow is generally designed for a velocity of 10 to 15 metre per second, and liquid lines for 2 to 4 metre per second flow. A mass of liquid travelling at five times the intended velocity can be a frightening thing.

If the pipe is not designed for slug flow, damage can occur, particularly at supports and elbows. Figure 7.1 shows the effect one such slugging event has had on a steam pipeline.

CASE HISTORY 7.15 Overflow Hammer [3]

Overflow of liquid into flare and vent lines can similarly give slug-flow hammer effects. An overflow of butane from an overfilled flare knockout drum at the Milford Haven Refinery in 1994 resulted in a rupture of the flare line at an elbow which had corroded. The butane was released and the resulting gas cloud exploded, destroying part of the refinery.

Operators need to know of the dangers of accumulating liquid and need to know how to judge the amount of liquid collecting in pipes. They especially need to understand the importance of draining off liquid which accumulates at low points and the importance of limiting gas flow velocities in piping in which there are both gas and liquid.

5. When liquefied gas is filled into a pipeline which has been shut down for a time but not cleared of gas, filling is often done with a closed pipe. As liquid is filled into the pipe, the gas is compressed. When the pressure rises high enough, the gas ahead of the liquid condenses.

   Great care is needed when this is done, to ensure that the last part of the filling is carried out very slowly. If filling is made at full speed, the gas bubble at the end of the pipe collapses very rapidly, and the hammer effect can cause pipe rupture.

In a safety engineering course, I asked the team of about 20 process and equipment engineers and plant supervisors how many knew about water hammer. About half indicated that they did. I then asked how many knew about liquefied gas hammer. Just two raised their hands. I asked where they had learned about it, because it is not widely known. They answered that it had happened to them two weeks previously on a 1-inch chlorine line with a considerable chlorine release (about 50 kilograms).

CASE HISTORY 7.16 Liquefied Gas Condensation Hammer [4]

A similar physical effect occurred at the Texas City Refinery in 1984. Liquefied petroleum gas (LPG) was run down from the LPG distillation plant to a storage vessel at the tank farm. Due to an instrument failure, the vessel overfilled, and the final collapse of the gas bubble caused a pressure spike. The safety valves could not respond rapidly enough. A section of the vessel about 1½ m in diameter, which had a weld defect, blew away. The LPG was released and ignited causing fires, and BLEVEs, at several other vessels [5].

Operators need to know that when filling anything with liquefied gas, there is a danger, and to do this in a safe manner, with a nitrogen cushion, with an open discharge valve or very slowly as the pipeline becomes filled. Operators, as always, need to understand the dangers of overfilling and that, especially in liquefied gas vessels, this can lead to catastrophic rupture.

The listing of momentum effects above is limited but detailed. The effects described are not generally known in their entirety, even to experienced process designers. From this limited list of cases it can be seen that the extent of knowledge of hazard potential needed by an operator or an operation supervisor is extremely large. Basically, all the accident effects which have occurred need to be reviewed, judged for relevance and learned. If this is left to the individual operator, there may be a few such effects which are picked up. A much more systematic approach is needed. This is currently a major gap in operations safety.

KNOWLEDGE-BASED OPERATING ERRORS

CASE HISTORY 7.17 Lack of Knowledge

After a shutdown, operators opened up a chlorine supply line to a reactor from a new chlorine drum. The supply valve to the 1-inch line was opened fully. During the shutdown, liquid chlorine, which had earlier filled the line, had evaporated and had passed into the reactor as gas. The line was therefore gas filled. When the supply valve was opened, liquid chlorine rushed into the line, compressing and condensing the gas. When the liquid hit the control valve, it ruptured due to the hammering effect.

In the investigation, it was found that neither the operators nor the plant engineers knew about this hammering effect (lack of knowledge of failure phenomena). Even if they had known, they did not know that the line was gas filled (lack of knowledge of actual plant status). If they had known that liquefied gas block or control valves must always be opened slowly, and only partially, until the line is known to be filled, the accident would have been avoided (lack of knowledge of good operating practice).

CASE HISTORY 7.18 Misteaching [6]

A team of firefighters had been taught about the dangers of BLEVEs from fires at propane and LPG storage vessels. They had especially been taught about the problems of projectiles being generated and being thrown, especially along the axis of the vessel. They were taught not to stand in line with the vessel axis.

A short time later at Herrig Farms, a fire occurred at a broken pipe on a LPG storage vessel. Firefighters were called and attempted to fight the fire. About 10 minutes after the firefighters had arrived, a BLEVE caused a large fireball (which is a more serious hazard to firefighters).

Erroneous learning, mislearning and erroneous teaching are fairly widespread occurrences which do not lead to accidents in many cases, though. As an example, most design engineers and operators taking part in HAZOP analysis training believe that pressure safety valves are very reliable and are the only pressure protection needed on a pressure vessel. (In fact, the failure rate when needed or the probability of failure on demand has been determined to be between 1 failure in 50 and 1 failure in 2, in different studies.) The reason that erroneous learning is a relatively rare cause of accidents is that knowledge-based thinking is only rarely needed in modern plant practice and that modern plants are heavily protected by automatic systems. Accidents due to lack of knowledge are much more common, especially in field operations and maintenance. Hydrogen sulphide accidents are among the most frequent causes of fatalities, for example, where persons enter sumps or other confined spaces and where others follow them for rescue and also become casualties. An example of incomplete knowledge is shown by the Three Mile Island accident (Table 9.1), where the operators concentrated on the hazards of overfilling the pressuriser vessel, therefore overpressuring the reactor. This led them (together with the misdirection arising from a failed level sensor) to overlook the hazard of a low level.

CASE HISTORY 7.19 Failure to Remember or Understand the Relevance of Knowledge

A plant operations manager said, ‘When I was the manager of the water treatment plant, I used to curse the operators that drained down oil into the oily water drain system without telling us first. Now I forget that all the time’.

Forgetting is a natural process when there is no periodic reinforcement of knowledge. Inability to recall knowledge is natural if there are other activities which demand full-time attention.

There is another phenomenon which has been observed by several investigators. If a person is trying to diagnose a situation, the first explanation which arrived at, and which appears to fit the evidence, will tend to dominate all further considerations. This is true, even if further evidence contradicts the first explanation.

This effect is reinforced within a group, where the first good explanation tends to suppress all later explanations.

PREDICTING KNOWLEDGE-BASED ERRORS

When attempting to analyse or predict knowledge-based errors, one is faced with a fundamental problem. Knowledge-based thinking is needed in order to be able to deal with unanticipated problems being unanticipated; it is likely that they will be difficult to anticipate. The analyst is faced with the need to identify unanticipated problems; then to determine the different kinds of knowledge needed to diagnose and deal with the problem and then to identify all the kinds of errors which could arise.

Even with all the improvements which have been made in analysis and predictive techniques over the years, this challenge seems beyond our current capability. Even more, it seems unlikely that probability and risk could ever be determined. There are just too many possibilities and too few occurrences of each to be able to form a proper statistical basis. Possibly something can be done by working with training simulators used in problem diagnosis training.

What definitely can be done is to look for typical patterns of knowledge-based errors.

Given that it seems impossible to analyse risk for knowledge-based actions, what can we do? The answer is similar to that of dealing with design errors—it is better to eliminate as many as possible, rather than trying to predict them and calculate the frequency. Some ways of doing this are as follows:

- Teach systematically the range of knowledge required by operators. Explain especially how things work, how they behave in unusual circumstances and how things fail. Note that simulator training does this to some extent, but that current simulator training is too limited in scope and covers one level of the knowledge needed, to be able to deal with the range of problems.
- Ensure that the knowledge about plant configuration is correct and up to date by means of good as-built documentation through management of change and proper distribution of updates.
- Train operators in good diagnostic practice, including investigation of all relevant input, rather than focusing on a single indication.
- Train operators in how to distinguish the signal from failed instruments and provide instrument self-diagnostics. (HART-type diagnostics are standard on many installations today [7].)

REFERENCES

1. J. R. Taylor, Industrial Accident Physics, http://www.itsa.dk, 2014.
2. A. B. de Souza Jr., E. L. La Rovere, S. B. Schaffel, and J. B. Mariano, Contingency Planning for Oil Spill Accidents in Brazil, Freshwater Spills Symposium 2002, 2002.
3. U.K. Health and Safety Executive, The Explosion and Fires at the Texaco Refinery, Milford Haven, London, 24 July 1994, 1997.
4. Marsh, *The 100 Largest Losses 1972–2011*, http://usa.marsh.com/Portals/9/Documents/100_Largest_Losses2011.pdf, Downloaded 2015.
5. J. A. Davenport, Hazards and Protection of Pressure Storage of Liquefied Petroleum Gases, in 5th International Symposium on Loss Prevention and Safety Promotion in the Process Industries, Canne, Societe de Chimie Industrielle, 1986.
6. U.S. Chemical Safety Board, *Investigation Report: Propane Tank Explosion*, 1998.
7. HART Communication Foundation, http://en.hartcomm.org.

8 Errors in Supervisory Control

A large part of the work of an operator of a process plant, a power plant or a pipeline traffic system involves monitoring the state of the system, detecting significant changes and responding to them. These activities are known as supervisory control.

There are several different styles of supervisory control and different activities.

In some cases, monitoring is a continuous task, typically involving continuously watching a large screen which shows a representation of a system. This is typical of systems monitoring traffic situations, such as vessel traffic services (VTSs) monitoring sea-lanes with heavy traffic and surveillance systems for air traffic and for military air threats. Such monitoring is performed occasionally in process plants, but it is in fact quite rare for a process plant to require continuous focussed monitoring.

One of the features of continuous monitoring is that the operator quickly builds up a mental model of the system being controlled. Evidence is the ability to recognise aspects of the controlled system and the ability to make predictions.

An example of this is from a VTS system using raw (analogue) data displays. Operators could recognise different types of ship by their 'signatures' and could predict potential collisions and forthcoming need for avoidance manoeuvres. Later, introduction of processed radar with automated tracking replaced analogue-form signatures with symbols. The recognition of symbols as ships remained, but types could not now be identified.

For a process plant, it is more common for monitoring to be cyclic. The operator(s) makes a review of the state of the plant every few minutes. The actual cycle time depends generally on time constants of the plant. If significant changes can occur in seconds and if retrievable consequences require control in less than 1 minute, monitoring is generally continuous. With long time constants, the operator examines periodically.

Modern control system displays are often (generally) made on video display. The usual number of displays is from two to four per operator. The available display space is only rarely adequate for complete supervisory control, and display designs are a compromise, with the ability to switch between different display windows. Control engineers often solve this compromise by providing a summary display, which is visible all the time. Operators solve this problem by cycling through detailed displays. Older plant designers solve the problem by providing large mimic displays as hardware on the wall. These provide a continuous indication of the critical features of the plant state. The newest control rooms solve the problem by having a large summary display projected or back projected on the control room wall.

The workload in monitoring is reduced by providing annunciation messages or signals for information concerning operational changes of state and alarms for information concerning safety or reliability changes of state. The supervisory strategy in this case often changes, with the operator responding to signals and alarms and only occasionally surveying indicators. Also, when annunciation signals are used as a primary tool for human–machine communication, the operator will often select display of just a few key process variables (such as pressure and flow rate for a boiler), which allows a quick-glance review of the state of the plant. The operator can then devote a large part of the time to writing reports, planning or even reading a newspaper.

In all cases, the success of supervisory control will depend on the quality of the representation at the human–machine interface. If some aspect of systems operation is not displayed, or cannot be displayed, supervisory control will be less than perfect.

CASE HISTORY 8.1 Loss of Attention

In an environmental protection system, neither the temperature of flue gas nor the flow rate of activated carbon could be measured, only the activated carbon screw conveyor motor current. Supervision was not possible from the control room. Fires occurred frequently. It became necessary for an operator to inspect the equipment every half hour. (Later a proper monitoring system was fitted.)

Not all aspects of performance of a system can be summarised by means of alarms and annunciators. Some variables are very difficult to measure, e.g. level in a stirred reactor, viscosity and granularity in powder flows.

In other cases, measurements are available, but abnormalities involve not just a single variable, e.g. mass balance in flow to and from a tank and feed and bleed in control of pressure in a shut-down pressurised water reactor. (Advanced supervisory control systems, such as mass balance–monitoring systems, provide a solution for this.)

ERRORS IN SUPERVISORY CONTROL—CHECKING VALUES

Temperature and flow rates must generally be checked as part of procedures. These values may be displayed on a dial gauge, as a number on a mimic diagram, as a value on a computer screen linear gauge or, in some cases, as a trend curve.

Checking involves first remembering that the check is to be made, then seeking out and identifying the correct display, checking the value against a remembered or displayed limit and acting on the result. In some cases, it involves relating two values together, e.g. comparing two temperatures.

The main causes of omission are simple forgetting, or distraction. The checking task is not one which is essential for achieving the overall goal, and performance of the task is therefore not self-reinforcing. Another possibility if checking takes time is that the operators deliberately omit the check. This can occur especially in the field, if they have to walk to a specific gauge and if they have an expectation that the valve will be in the correct position. The main cause of selecting the wrong variable to look at is that the variables look similar or have similar labels or appear on similar displays. In the control room, checking may be omitted if the check requires changing screens, especially if the check rarely shows a problem, so that the expectation is that the check is OK.

Checking the value requires that the operator can remember or work out the correct value. There are several possible causes of error, including misremembering the threshold value or misremembering the indicator or item of equipment to which the value applies.

In some cases, display coding can be used, which allows the operator to judge the level of a hazard in a value without needing to remember target or hazard threshold values. Some observations of value-checking operations from a chemical reactor operations simulator study are the following:

1. Observation of cooling water temperature before starting a chemical reactor: Starting in summer, too high a temperature could be hazardous. The high summer temperature served to remind the operators, but omitting the check rarely had any adverse consequences. Observation over a period showed the check to be omitted 17 times in 200 start-ups. The check involved going outside to the cooling water tank. (This system was replaced at the end of the study.)
2. Filling day tanks from storage tanks in a refinery: About 10 transfers were made per day. In the course of a year, pump trips due to insufficient feed (too low storage) occurred 12 times, giving a probability of error of 0.00034 per operation. This is a lower limit on the number of checking mistakes and omissions, since such errors could become apparent only when storage tank levels were too low about once in five cases, giving estimated probability of 0.017 per operation.
3. On a feed water treatment to a boiler, checking *p*H, which required selecting a new display screen, was observed to be omitted 32 times out of 970 start-ups. (The results were taken from control system records. The display was redesigned after this study.)
4. Simulator studies were made for a batch reactor production and for a boiler start-up. The current operational records for the simulators of batch reactor production and boiler start-up are 60 and 342 operations sequences, respectively. Problems were inserted into all of the simulations, including critical flow rate checks. Oversights or erroneous checks were found for respectively 3 and 22 simulation sessions, giving error probabilities of 0.05 and 0.064, respectively. Follow-up interviews were obtained for the simulator sessions.

CHECKING A COMPLEX PRECONDITION

Checks of materials compatibility, temperature compatibility and system status often require a level of knowledge which goes beyond the procedural.

CASE HISTORY 8.2 Checking Complex Conditions

An operator of a catalytic cracker had operational problems, because the bottoms (hot heavy oil) from a distillation column could not be sent to the heavy fuel oil tank. He found an empty tank, which could be used by opening a number of manifold valves, and decided to use it. Unfortunately, the tank contained a residue of gasoline, which boiled when the hot heavy fuel oil was admitted. The roof of the tank blew off (and covered a car park with heavy oil residue; no one was hurt).

TABLE 8.1
Performance in Monitoring for Some Operations

Case	Operation with No Preinformation	Operation with Operator Preinformed	Errors with No Prewarning	Errors Preinformed
Tank transfer	100	100	14	2
Batch reactor	30	50	2	1
Compressor start-up	20	20	2	0

Checking complex preconditions such as the potential of interaction between hot oil and gasoline, as in the case above, requires several kinds of knowledge about the general physics of processes. (This is not necessarily formal knowledge. In many ways, knowledge from experience is better.) It requires knowledge about the structure and general functioning of the plant. And it requires knowledge about the specific state of the plant. It also requires the ability to process all this knowledge and integrate it, often implicitly and intuitively.

In Case History 8.2, the operator knew about the danger in general of mixing hot oil and gasoline but did not remember that there could be a small residue in the tank. In his mental model, an empty tank was truly empty. (This was elicited in the subsequent inquiry.) The belief was reinforced because the tank level gauge read zero.

It is extremely difficult to obtain statistical evidence for this kind of 'error' from actual plant operations unless the checking is mentioned implicitly in the operating procedure or in standard operating practices; it is perhaps unreasonable to expect an operator to make the connections needed. In fact, many risk analysts overlook the possibility of the example described above when making analyses.

A number of simulations were carried out of the kind described above. These included mass transfer between tanks in a storage terminal a simulation of a chemical batch reactor, and a compressor start-up. A number of complex failure conditions were inserted into the simulation. The simulations were carried out with several runs per operator. In some cases, the check was included into the procedure. The overall error rate is tabulated in Table 8.1.

The error rates depend on the complexity of the conditions, the way information is presented and the degree of experience of the operator.

From observations in the control room, a very experienced operator seems able to check for unusual conditions more times than not. A list of 20 surprising situations and corresponding very competent operator actions is given in Ref. [1]. From simulations, for operators in general, the following error rates seem consistent with observations: 0.1 for observing a complex precondition not described in the procedures and 0.03 when the precondition is described in the procedure, in both cases given a good display and training and that the operators actually read the procedure.

RESPONDING TO ABNORMAL CONDITIONS

The signal which the operator perceives from the display system is called the *cue* for action. This can be an annunciation signal or alarm message, or it can be recognition of a need for action on the basis of display, such as a parameter value or a trend curve.

When a need for response is identified, there is a need to associate the cue to an action. This may be a simple association to a standard action, e.g. when tank level is high, shut down the transfer pump (rule-based activity; see Chapter 5). Supervisory control, though, often involves diagnostic activity. E.g. pump is indicated as running, but flow rate is low. Is the filter blocked? Is the liquid viscosity too high (temperature too low?).

There are situations, however, where operations and production must proceed with known abnormal conditions. Examples are production with inadequate steam supply due to a boiler failure, operation with inadequate heat exchange due to heat exchanger fouling or operations during very cold weather. In these cases the plant performance will be away from normal, and the operator needs to adapt. The associations, control targets for parameters and the responses to alarms may be different from those under normal operating conditions. For example, cooling failure in a chemical reactor would normally be indicated by a high-temperature alarm, but improvised cooling using a potable water supply as source may not be fitted with instrumentation at all.

Many of the abnormal conditions are quite standard, and some may even be predictable by the operator. Filter blockage, for example, is expected on a routine basis. It is very rare, however, for written operating procedures to take abnormal operation modes into account.

CASE HISTORY 8.3 Problem Anticipation from Memory

During studies in the control room of an oil terminal, the operator explained to two of us that he would have to go out into the field to adjust a pump. On arrival at the pump, we found that it had a very long suction line (1/2 kilometre), which is normally considered poor design. As oil was transferred from the source tank, the level in the tank would fall, and with the flow resistance in the long suction line, the pressure at the pump suction would become too low. The lack of sufficient positive suction head would cause cavitation, causing the pump casing to erode quite quickly. The operator closed the pump discharge valve slightly, so that the flow would be reduced and, with it, the pressure in the suction line would increase. The operator knew from experience just when to do this, so that full pumping rate could continue as long as possible while preventing damage to the pump.

OPERATIONS WITH FAILED INSTRUMENTS

An important aspect of supervisory control is that it is generally very dependent on instruments and sensors. The instruments are not perfectly reliable, and on a large plant, there will always be some instruments in a failed state. In some plants audited, as many as 30% of the instruments were out of service, awaiting repair.

Ideally, operators need to know how instruments work and how they fail. They must know how to continue operating the plant with some of their instruments 'blind'. Very often, this requires that operators go out into the plant simply as an observer. Ideally this is the first or second field operator, with radio communication back to the board operator. However, it is not unknown for single operators to have to go into the field, leaving the control workstation unattended.

One of the problems with this kind of practice is that it introduces a can-do attitude, where the plant is kept running at all costs. There are some instruments which are absolutely essential to safe operation, for example, temperature trips on a reactor with an exothermic reaction and runaway possibility. All plants should have minimum conditions for permitted operation, and proper function of safety systems and critical instruments should be among these.

IMPROVISATION AND OPERATION UNDER ABNORMAL CONDITIONS

It is often necessary to operate plants under abnormal conditions. If a particular tank is leaking, for example, it must be shut down and repaired. If there are alternative tanks, it will be possible to continue production, with an alternative tank to receive product and with a different lineup of manifold valves.

The range of such abnormal operation is very wide. Some examples I have observed are the following:

- Continued operation of a plant with a small leak of kerosene on a pump: The leak presented a hazard for fire and a larger hazard for domino effects of fire growth. A fire watch was posted and operators were required to watch out for ignition, and foam monitors were trained on the pump.
- A reactor was operated for a relatively long period with overheating of the vessel due to failure of the internal insulation. Water was sprayed on the vessel to keep it cool. Operators were required to watch for any change in conditions and to be especially cautious about temperature transients.
- A plant could not achieve adequate net positive suction head in a pump after a revamp, due to high temperature in the feed. Until a new cooler could be installed, it was necessary to continually drench the piping and pump with water externally. Operators were required to monitor temperature. The heat exchanger corroded externally to an unusable condition quite quickly.
- A steam turbine was operated for a relatively long period with a crack in the turbine housing. Clamps were applied to prevent catastrophic failure, and the rate of production reduced.
- A particularly toxic substance was to be processed. Operators not directly involved were required to evacuate the production area, and no one was to enter during the production campaign unless suitably suited and masked.
- During a hot weather period, adequate chemical reactor cooling could not be obtained, due to poor cooling tower performance, with an especially high risk of reactor runaway and limited emergency cooling. Operators were required to monitor temperature continuously, on a minute-to-minute

basis, rather than on the '5- to 10-minute plus end of batch' basis, which was more normal.

- During unscheduled maintenance of a pump, a plant was operated with only one train out of two in one stage and at half capacity in all other stages.
- During operation of a power distribution network, if some generation equipment is out of operation, the network may be operated very close to overload, and operators are required to shed load. Which loads to shed presents a problem unless there is a predefined load-shedding priority list. Even if there is, there can be problems to be taken into account, such as not shutting off power during a tank truck loading unless transfer lines can be emptied.
- During a flood in a refinery, operators were required to keep boilers and power generation working 'at all costs', since power was needed to keep steam turbine–driven water pumps working. This required a lot of improvisation since fuel supplies and steam lines were cooled by the floodwater, causing blockages and condensation, and some pumps had to be shut down since they could not operate under water.
- A frequent situation is operation of a plant with an out-of-date mimic diagram in the control room, or even on the control computer.
- A frequent situation arises when some instrumentation is out of operation. In some cases, interlocks may be bypassed.
- A decision was made to rework a contaminated batch of a nitrated product by distillation. The viscosity and concentration of the already processed product were too high, however, and a runaway and an explosion occurred.

The problem for operators working under abnormal conditions is that they must continually remember undocumented aspects of plant status and make allowances. The abnormal conditions may increase risk, as in Figure 8.1, where small disturbances could lead to overheating of a bearing and to fire. Risks may be analysed

FIGURE 8.1 Operation of a generator with overheated bearing. An air mover is used to provide additional cooling while waiting for delivery of new parts.

for these abnormal conditions and job safety analyses performed, but this is not always done (the standard of 'engineering' in Figure 8.1 is awful and could surely be improved on with a few hours' work by any reasonably competent maintenance crew). Even after the analysis though, there are problems of remembering the abnormalities, especially in an emergency, and problems in communicating the abnormal conditions between different shifts.

In modern safety practice, the range of abnormal operating situations allowed is specified in a manual (or matrix) of permitted operations (MOPO). This specifies the 'envelope' in which it is permitted to operate the plant, including the following:

- Pressure and temperature limits
- Limits for critical supplies, such as lubricating oil and firefighting foam
- Limits for extreme weather under which the plant may operate
- Limits on safety system outages, such as fire water pumps and shutdown devices
- Limits on the exposure of persons and on occupancy, for example, ensuring that there is sufficient lifeboat capacity on a platform
- Limits on the procedures which may be carried out, and which may be performed in parallel

It is important that the MOPO specifies the permitted operations, rather than giving a list of prohibited operations, in order to avoid the assumption that anything not prohibited is allowed. Safety assessments should be made for all MOPO-permitted limits and operations.

SHUTDOWN DECISIONS

In modern practice, all serious disturbances in a process plant should be subject to instrumented monitoring and automatic trips. Hazard studies are made to ensure that the emergency shutdown system is adequate, based on risk studies. In the best of such studies, no credit is even taken for the operator to make the shutdown decision. The job of the operator is to correct the disturbance before shutdown is needed.

(There is an exception to this rule—shutdown in response to fire and gas alarms is still generally made manually because of the relatively high degree of unreliability of these systems, and the unspecific nature of the indications they give. A fire detector, for example, will indicate where a fire is but not the source of the flammable material. In the best modern systems, instrumented fire detection is supported by closed-circuit television to help the operator define the type of shutdown needed.)

Abnormal operations are only rarely subject to detailed hazard analysis, but they are quite often subject to careful judgement by the operations manager. Such operations can be performed in a fully responsible way, or irresponsibly, depending on degree of risk. While it is easy to say that safety regulations should be followed, many situations will always arise which are not covered by regulations, or are borderline cases.

In fact, the decision to shut down a plant because of hazard is always a difficult one unless the automatic system forces shutdown. Operations in which shutdown is not automatic are often borderline cases. In large plants the cost of shutdown can be several million dollars. Also, shutdown itself can sometimes worsen the problem. For example, when a vessel is overheating, shutdown can sometimes cause a pressure transient and activation of blowdown systems, exacerbating problems.

OPERATOR PERFORMANCE UNDER ABNORMAL CONDITIONS

During normal operations of a modern plant, the actions to be carried out by the operator are often fairly simple. In a fully automated plant, all he or she needs to do is monitor operations, check that nothing abnormal is happening and, if anything abnormal does happen, try to find out why. The actual demands on the operator in such cases can be very limited, even to the extent that there is little to retain the operator's attention.

During abnormal operations, operators are required to operate according to non-standard procedures. They must adapt to the new procedures. Often, a higher degree of vigilance, as well as a faster response to disturbances, is required. Also, the new procedures may be communicated verbally rather than in writing. The operators may note down the modified procedure during verbal presentation, introducing a new potential for error, or they may try to remember, introducing a greater possibility for error.

If the abnormal operation continues for a long period, operators may have difficulty in remembering the abnormal plant state. This applies especially if the abnormality concerns secondary systems such as lubricating or seal oil systems or seldom-visited systems. Problems are almost guaranteed if operating staff are transferred to a unit which is in an abnormal state, unless full and detailed briefing is given, and there will be errors if some details are forgotten.

CASE HISTORY 8.4 Abnormal Plant Condition

A nuclear reactor vessel was ruined when tested due to seawater entering via a pipe earlier installed for testing a steam condenser and had not been removed. No one had remembered the improvised piping.

One special problem is that abnormal operations may be performed on the basis of an incorrect theory of operation, or a false assumption. For example, one refinery was operated for a period with one flare line shut down for modification, and all vents bypassed to a second line. Subsequent review showed that back pressures involved in the improvised setup had rendered some safety valves inoperable, although luckily they were not required to activate during this period.

Another example was the operation of a plant with two out of three cooling pumps out of operation. This was considered acceptable. On review, the single cooling pump was found to be adequate for average cooling demand but not for peak cooling demand.

One of the problems in this kind of situation is that operation engineers must make calculations and judgement, but the full plant design rationale is only seldom available. For units which are provided by process licensors, telefax calls to the design team can sometimes give greater security, but such confirmation is sought only in cases where operation engineers recognise that there is a problem.

A further feature of abnormal operations is that to develop the revised procedure, reference must be made to plant drawings. If drawings are incorrect or out of date, abnormal operations may contain traps for the unwary operator.

OPERATOR ERRORS IN ABNORMAL SITUATIONS

One of the main problems of operations under abnormal conditions is that by definition, safety systems will not be designed for such operation. Safety then becomes much more dependent on the operator.

A range of operator error types is associated with abnormal operations:

- The operator may not know about the abnormal operation. This arises if the situation is not explained, some detail is forgotten or an operator is transferred from another unit or returns from holiday without briefing.
- The operator may forget the abnormal operation and proceed as normal.
- The requirement of abnormal operation may place too high a demand on the operator for precision in operation or for response time. At least, the probability of error will be increased.
- The abnormal operating method may in fact be erroneous. Particular problems are not checking for necessary preconditions for operation or possible side effects or correctness of assumptions.
- It may be impossible to react to disturbances, simply because the system does not have adequate disturbance control capacity.
- Novel or unexpected disturbances may arise, requiring improvisation, often of a type not anticipated or subject to earlier safety analysis. Note that virtually all safety analyses are based on normal operations. Mistakes are easily made under such situations.

In modern practice, most of these problems *should* be prevented. No change should be made to operations without being subject to management of change (MOC) review. For most of the changes described above, this would require a properly written procedure and a mini–HAZOP study or a job safety assessment. Checking improvisations in this way is in fact the situation in all high-integrity organisations today. There are many other organisations for which MOC regulations are breached.

Even for those companies which follow regulations strictly, there are problems, however. When improvisation is required in order to get production restarted, there is considerable time pressure (see the case histories in Chapters 7 and 12). When improvisation is necessary to ensure safe shutdown, there is even greater time pressure. During this time, the most experienced operators and operation engineers are needed for controlling the problem, and the last thing they need is a demand to sit in a HAZOP team meeting for several hours. Preparation may take time, with drawings needing to be found. Design engineers are unlikely to be readily available. Even finding a good HAZOP team leader may be difficult.

REDUCING ERROR IN ABNORMAL SITUATIONS

Some of the problems can be reduced if good practice is followed, so that the teams are prepared for a rapid response:

- Good as-built drawings should be kept up to date, if necessary with red marking of changes which have not been updated at the drawing office. Software is available today which allows this to be done in a well-controlled and fully documented manner.
- Drawings should be available to the operations staff, with a reasonably convenient printer.
- MOC review should be carried out for all improvised operations.
- There should be at least a couple of persons in the company with training as HAZOP team leaders. It is usual today to rely on third-party HAZOP leaders, but this is seldom practical under time pressure.
- Good safety philosophy documents should be available for the plant, so that the team at least knows which principles it is breaking when it improvising and is then able to compensate for these.

REFERENCE

1. J. R. Taylor, A Catalogue of Human Error Analyses for Oil, Gas and Chemical Plants, http://www.itsa.dk, 2015.

9 Emergency Response and Disturbance Control

CASE HISTORY 9.1 Rapid Emergency Response in a Complex Situation

My friend Yos and I were quietly observing control room operations for a large ethylene plant, from a desk at the back of the control room. Suddenly, alarm bells began to ring, and the control panels began to blink red. The supervisors (two of them) jumped up and stationed themselves behind the panel operators (response time: about 10 seconds). About 30 seconds into the incident, one of the supervisors shouted, 'It's a dip'. This means that the electrical power supply had lost voltage for a short time and had then recovered. Later, it would prove that the electrical power lines for the area had been struck by lightning, and the 'dip' was the time taken for switch gear to close to provide an alternative supply.

The dip had caused automatic cutout relays to shut down pumps (and compressors, but these are so massive that they keep running for 2 to 3 minutes).

A wave of understanding passed across the row of faces of the operators. Their task was now clear—restarts need to be made, to get pumps operating again (about 80) before receiver vessels overfilled, or feed vessels emptied. Ideally, the restarts occur in the proper sequence, with the most acute situations being dealt with first, but each operator has to set his own priorities. There are procedures for 'black start' and procedures for 'start up from standby', but there were no procedures for 'start up from a ragged partial shutdown'.

The operators got the plant stabilized gradually over the next 30 minutes. The supervisors' quick recognition and the operators' capability meant that total shutdown was avoided, saving about $1.3 million in lost production. The cost of emergency shutdown (ESD) is so large because restart requires checks, slow start-up of furnaces, a lot of work in the sequence of operations and careful reestablishment of conditions in distillation columns.

In contrast, the neighbouring plants on the same main power supply suffered total ESD. Local in-plant power stations were unable to take up the load fast enough to prevent the pump trips, and the ripple effect of these through the plants caused individual unit trips and losses of power.

The case shows the importance of the ability to see and understand a complex signal (literally hundreds of red alarm lights) and to diagnose the problem. It shows also the importance of the instant understanding of the code word ('It's a dip') on the part of the operators and what they should do—this level of response is almost impossible to achieve without earlier experience of something similar.

The supervisors' recognition of the situation was at a high level. Interviewed later, he said that he had seen similar but less extensive situations earlier. Also, everyone knew that with a short reduction in power supply voltage, pumps would trip and needed to be restarted and flows needed to be controlled to ensure mass balance. Heating balance was carried out more or less automatically. Control system power was maintained by uninterruptible power supplies, so there was no real loss of control.

The response shows that each operator understood, i.e. had a clear mental model, of the mass balance and the importance of pump operation on this, as well as a clear understanding of the word *dip*. The fact that other plants in the area did not keep operating was in some cases the result of greater time sensitivity (less product buffering between production stages) but mostly because the cause of the incident was not recognised fast enough.

The importance of this kind of expertise in emergency response is illustrated by other much less fortunate incidents, described in the following section.

EMERGENCY RESPONSE

Plant operators have several tasks in an emergency. Modern plants have ESD systems which should prevent disturbances from turning into accidents. Examples of emergencies where operator action is definitely required are the following:

- Activation of fire-suppression system: Very few onshore fire-protection systems such as deluge or fixed fire water monitors are activated automatically. Manual activation is used in order to reduce the number of spurious activations. Manual activation is also used in order to concentrate the use of fire water to the location of the fire.
- Operator action in the field is required to close off releases from large inventories, where there is no ESD for inventory isolation, where ESD closure of valves fails or where the inventory between two ESD valves is large.
- Manual operation is generally needed for plant unit depressurisation.
- Most oil, gas and chemical plants do not need long periods of shutdown operation in ESD state, as does a nuclear power plant, in order to ensure continued reactor fuel replacement. There are often long periods, though, where operators struggle to keep a plant operating. (See, for example, the case of the Milford Haven Refinery explosion in Table 9.1.) In a large fire though, operators will often struggle to reduce pressures, to transfer inventories and to adapt to losses of power. Such actions require understanding of the plant itself, understanding of prefire plans and the ability to improvise correctly in the cases where no preplanning has been made.

TABLE 9.1
Features of Some Large Accidents

Accident	Description	Faulty Instruments	Inadequate Instrument Design	Poor Control HMI	Poor Procedure	Lack of Belief in Procedures	Inadequate Training	Inadequate Knowledge
Three Mile Island, 1979 [1,2]	A safety valve opened and failed to reclose. A level indicator on a pressuriser vessel failed stuck. Operators afraid of overfilling reduced cooling water flow until the reactor core was uncovered and a partial meltdown occurred.	Level sensors failed	Poor instrumentation for safety valve leakage		Emphasis on overfilling, little on low water level		Yes, emphasis was on overfilling, not low reactor level	Incomplete knowledge
Milford Haven Refinery, 1994 [3]	After a thunderstorm with many compressor trips, a level indicator on a debutaniser feed vessel failed. Level rose and butane flowed to the flare header. The flare knockout drum overfilled, and a slug of liquid butane flowed to the flare. Liquid hammer ruptured the flare line, butane released and a vapour explosion occurred.	Level sensor stuck; many trips due to lightning	Insufficient instruments to be able to determine mass balance	Yes, difficult to see an overall picture and not possible to diagnose mass balance problem	Yes, emphasis on low level, not on overfilling		Yes, this accident led to a major boost in the use of simulator training	No situational understanding

(*Continued*)

TABLE 9.1 (CONTINUED)
Features of Some Large Accidents

Accident	Description	Faulty Instruments	Inadequate Instrument Design	Poor Control HMI	Poor Procedure	Lack of Belief in Procedures	Inadequate Training	Inadequate Knowledge
Texas City, 2005 [4,5]	During a post–turn-round start-up, a distillation column outlet valve was left closed by mistake. One level sensor failed. A second level sensor did not show true level. The column overfilled, and naphtha flowed to the vent knockout drum. Naphtha flowed from the knockout vent drum ignited and exploded.	Level sensor failed, not repaired	Second-level sensor alarm set points were frequently exceeded during start-up; rangeability of second sensor inadequate		Procedures were out of date, therefore not used			
Avon Refinery in Martinez, 1997 [6]	A difficult operational condition arose as a result of a flange leakage and the need to take one of three hydrocracker trains out of operation. The load was distributed to the remaining two trains. A hot spot arose in a reactor bed. Efforts to control the temperatures persisted too long.	Yes, history of failure of a new intelligent alarm system	Yes, default condition for out-of-range high level was zero. Readings bounced between zero and high.	Several systems, with instrument inconsistencies; needed instruments readings not available in control room	Yes, procedures out of date and inconsistent, not used by operators	Yes, procedures were known to be wrong; operators failed to follow explicit emergency procedures	Yes, although there had been a recent course in hydrocracker safety	Erroneous knowledge

(Continued)

TABLE 9.1 (CONTINUED)
Features of Some Large Accidents

Accident	Description	Faulty Instruments	Inadequate Instrument Design	Poor Control HMI	Poor Procedure	Lack of Belief in Procedures	Inadequate Training	Inadequate Knowledge
Ammonium nitrate plant, Port Neal, 1994 [7]	Ammonium nitrate unit was operated at very low *p*H. Earlier failures had resulted in contaminations with chlorides and possibly oil. 200 psig[a] steam was used to clear a nitric acid sparger. The ammonium nitrate exploded and many were killed.	Yes, failed *p*H sensor, one of the most important instruments in the unit; not replaced because spares were out of stock	The unit was very much under-instrumented by modern standards. A SIL study would be required today and would result in much more thorough monitoring. *p*H sensor impulse lines were not insulated or heat traced. There was a delay of 2–3 h in *p*H registration due to sample point location.		Yes		Yes, not trained in the operator training or refresher courses	Apparently the entire company unaware of the hazards of sensitisation of ammonium nitrate by chlorides, low *p*H or bubbles from injected steam or air

[a] psig: pounds per square inch gauge.

ERRORS IN EMERGENCY RESPONSE

The most crucial aspects of emergency response are those which result from errors. The simplest errors are failures to act.

The tragedy at Bhopal is an example. Water leaking into a methyl isocyanate (MIC) tank resulted in a runaway reaction and a spray of MIC which ultimately killed at least 8000 people and injured many tens of thousands. During the accident, which lasted many hours, alarm was not given to the authorities or emergency services, even though the operators knew fully well what was happening at the plant itself. Why not?

CASE HISTORY 9.2 Error in Emergency Response—Increased Pumping When a Leak Occurred

There are many cases of erroneous response to pipeline ruptures, in which operators compensate for the loss in pressure by increasing flow. The Cubatão fire, in Brazil, involved rupture due to hammer in a 10-inch pipeline. Pressure signals from the receiving station, about 80 kilometres away, resulted in the operators increasing pumping. The kerosene flowed down a stream, to a village, ignited and killed an estimated 850 people.

CASE HISTORY 9.3 Error in Emergency Response—Another Increase in Pumping

A similar incident, at Ufa, in Russia, occurred in an LPG pipeline. The LPG continued to be pumped at an increased rate and gave a gas cloud several kilometres in length. The cloud was ignited when two trains entered it.

CASE HISTORY 9.4 Error in Emergency Response—Yet Another Increase in Pumping

A similar case investigated by the author sheds some light on the thinking behind such incidents. The case involved the simple rupture of a valve in a gasoline transfer pipe to an export terminal. The operators noted the loss in pressure and therefore increased the pumping rate. About 20,000 barrels of gasoline were lost, with no ignition and no injury. On interview, the operators explained that low discharge pressure often resulted from changes in viscosity and required additional pumping effort. This was actually not the case for gasoline but was, and is, definitely the case for other heavier oil products. The operators did not have sufficient knowledge to differentiate between viscosity changes of fuel oil and the almost constant viscosity of gasoline.

Once again, the problem arises from a conflict between the day-to-day operational goals, the requirements of emergency response; the lack of any instrumentation which clearly indicates an emergency condition, and the lack of imagination or knowledge on the part of the operator. The problem could be solved in all of the cases described by a simple mass balance leak–detection system, at almost insignificant cost.

Much more difficult emergencies can arise.

CASE HISTORY 9.5 A Prolonged Struggle to Control an Emergency [6]

An explosion and following fire occurred at a refinery, resulting in 1 fatality and 46 injuries. The accident resulted from a runaway reaction in a hydrocracker, causing a high temperature, failure of the vessel discharge piping and release of a mixture of hydrogen and hydrocarbons, which ignited and exploded.

A hydrocracker is a vessel in which hot hydrocarbons are reacted with hydrogen in the presence of beds of catalysts. The reaction generates heat, and a temperature excursion occurred due to an uneven flow through the catalyst beds and a hot spot.

The catalyst beds were heavily instrumented with thermocouples, so that temperatures could be thoroughly monitored. Measurements were also recorded on a data logger, and alarms were set at 780°F. The maximum operating temperature was 800°F. Above this temperature, it was known that a runaway was possible.

An emergency depressuring system was provided, and it was also possible to divert the feed to storage. Additionally, it was possible to provide a quench flow of cooled hydrogen to each bed to control temperature.

The sequence of events leading up to the accident was as follows:

- January 10: The temperature monitoring was switched from the data logger to the distributed control system (DCS).
- January 14: A temperature excursion occurred, with a temperature up to 900°F, on bed 4 of reactor 1. Temperature was controlled, but operators reported problems with the 1/A temperature-monitoring system.
- January 20: Use of the DCS was discontinued and the data logger was put back into operation.
- January 21, 4:50 AM: The reactor-discharged heat exchanger began to leak again. Some feed to reactor A was diverted to reactors B and C, causing reactor cooling and high nitrogen content in the gas flow.
- 10 AM: Some catalyst beds were poisoned by high nitrogen concentration resulting from the low throughput on reactor A.
- 7:34 PM: A temperature excursion occurred in reactor 3, bed 4, causing inlet temperature to bed 5 to increase rapidly.
- 7:35 PM: The quench valve to bed 5 was opened wide, bringing bed 5's temperature down. Data logger readings bounced from zero to normal to high and back. Bed 4's outlet temperature decreased to 637°F.
- 7:37 PM: Bed 5's temperature to bed 5 was increasing. The operator manually closed the bed 5 quench valve.
- 7:38 PM: The opener valve was reopened. Bed 5's outlet temperatures reached 1200°F.

- 7:39 PM: Operators received a message from a field operator which could not be understood.
- 7:40 PM: Bed 5's temperatures were off scale and defaulted on the logger to zero. The operators now had no detailed indication of temperatures. Operators requested the assistance of an instrument technician.
- 7:41 PM: One of bed 5's outlet temperatures read 1398°F on the data logger. A section of reactor 3 piping ruptured, causing an explosion and fire. The field operator was killed.

When this sequence of events is read, it is easy to see the operators struggling with control while at the same time being uncertain about the validity of instrument readings. It should be noted that quench was on manual control. If it had been on automatic control, it could have compensated for the high bed inlet temperature but not the runaway in the bed, due to the control loop design.

The closure of the quench valve appeared to be a mistake, quickly corrected by a second operator.

There was an emergency plan for the hydrocracker, actually posted on the control board. For temperature excursions, the actions to be taken were as follows:

- For temperatures 5°F above normal, change reactor controls to reduce temperature to normal, including reducing trim furnace firing, increasing compressor throughput or using quench hydrogen on individual beds.
- For temperatures 25°F above normal, close the hydrocarbon feed valve, reduce trim furnace firing, circulate maximum hydrogen and maintain pressure. Use quench on individual beds. Cool the reactor at 100°F per hour down to 500°F.
- For temperatures 50°F above normal, activate depressurisation, which also shuts down compressors and pumps and trips the furnace.

It is obvious with hindsight that this was a case requiring a shutdown of the system (the third response option). The very poor design and performance of the instrumentation, though, definitely left the operators in doubt. There is always significant management pressure when operating a plant of this kind to keep running as long as possible while trying to solve the problems, because any shutdown involves a large financial loss. Adding to the doubt were earlier temperature excursions where the shutdown required by procedures had not been made, with a successful control of the disturbance. No corrective action was taken by the management to emphasise the need to follow emergency procedures. Additionally, depressurisation had in the past caused liquid to be released from the flare, causing 'burning rain' and resulting in grass fires.

One further aspect of Case History 9.5 is important for the interpretation of the accident. Procedures were out of date, not being changed when catalyst effectiveness was upgraded. There was conflicting guidance between versions of the procedure, and bed temperature differential limits were not stated.

There are several other problems noted in the Environmental Protection Agency accident report, mostly concerning design and management error, rather than operator error.

Case History 9.5 is typical of several accidents where operators struggle to control disturbances. Factors which appear regularly are as follows:

- Failed instrumentation, so that the operators do not know the actual state of the system and are in doubt about the information they do have
- Poor instrumentation and display design, so that the operators were not informed or even misled
- Poor emergency procedures, with a lack of explicit criteria for action
- Earlier noncompliance with emergency procedures, without significant consequences
- Poor hazard analysis, with inadequate consequence analysis, leading to poor appreciation of risks
- Consequential poor training of operators about how to deal with disturbances

ANALYSIS FOR EMERGENCY RESPONSE

Analysis for emergency response involves, first and foremost, analysis of the full range of disturbances and emergencies which can arise in a process plant. A good hazard and operability analysis will provide a starting point but is not in itself sufficient. More dynamic analyses, such as the step-by-step analysis procedure described in Chapter 18, provide more information. Process calculations will need to be made for the most disturbances, to determine response times, and process simulation will be necessary for the more complex disturbances.

CASE HISTORY 9.6 Emergency Response against a Deadline

A natural gas-gathering pipeline could be depressurised readily during the first few days after a shutdown, but if the pipeline cooled down, depressurisation could cause methane hydrate formation and a potential for blockage. Blockage would require about a week to clear. A dynamic calculation was needed to determine at which times depressurisation would be a feasible emergency response.

Having identified a potential serious disturbance, the first step is to determine whether the operator has any chance of dealing with it in time. If not, then the disturbance must be dealt with by the automatic shutdown systems. The HAZOP analysis needs to be backed up by a description, preferably also drawings of how each emergency will appear to the operator. The indication should be clear and unambiguous and should give as much information as possible about the state of the plant and disturbance causes and potential consequences. The operator response can then be described. The operator response then needs to be analysed to ensure that there is sufficient time and that the response is reliable.

Chapter 18 describes a methodology which was developed for checking operating procedures for error. It includes facilities for checking emergency procedures, including rough calculations of 'grace time' (the time available before manual emergency response becomes ineffective) and estimates of reasonable expectations of operator response times, based on research into human factors. For complex and critical systems though, full dynamic calculations are needed, such as those which can be made with process design software or with special-purpose simulators such as those for plant depressuring and for pipeline operations.

ERRORS IN EMERGENCY RESPONSE

Errors in emergency response can be as follows:

- Simply overseeing or ignoring an indication or alarm (see 'Distraction', Chapter 3) due to the fact that the key problem is an extensive cascade of alarms: This is the 'alarm shower' problem (see 'Alarms and Trips', Chapter 21).
- Mistaking one situation for another, possibly due to ambiguity of the indications, possibly due to focussing on a subset of indicators or due to a straightforward mistake: Deeper-level causes are lack of training, erroneous training or wrong focus in training.
- Judgement errors in the extent of effects of emergency response, e.g. overestimating the effects of coding, dilution, etc.
- Lack of appreciation of the seriousness of the situation.
- Any of the standard list of errors (see Chapter 18) in carrying out the emergency procedure.
- It is possible to provide excellent training in emergency response by using hazard analysis–based simulators. The actual situation, though, is that present-day training simulators are often based on deficient hazard analyses, so that there is a danger that these contribute to accident potential in some areas, while preventing accidents in others.
- Error potential reduction for errors in emergency response.

REDUCING THE POTENTIAL FOR ERROR IN EMERGENCY RESPONSE

There are several steps which can be taken to reduce the potential for error in emergency response:

- Preincident planning for a full range of situations requiring emergency response: For process plant, a combination of HAZID and HAZOP studies is a good starting point for identification of potential emergency situations, but then it is necessary to describe each emergency scenario in detail and then to analyse the response for adequacy of resources and for timeliness [8]. The planning should be plant and location specific.
- Emergency response tabletop exercises.

- Emergency response field exercises.
- Emergency situation awareness training, preferably based on videos from actual emergencies, in order to acquire familiarity with the scale of possible emergency events.

REFERENCES

1. U.S. President's Commission on the Accident at Three Mile Island, *Report of the President's Commission on the Accident at Three Mile Island: The Need for Change: The Legacy of TMI*, New York: Pergamon Press, 1979.
2. M. Rogovin, *Three Mile Island: A Report to the Commissioners and to the Public*, Nuclear Regulatory Commission, Special Inquiry Group, 1980.
3. U.K. Health and Safety Executive, The Explosion and Fires at the Texaco Refinery, Milford Haven, London, 24th July 1994, http://www.hse.gov.uk.
4. U.S. Chemical Safety Board, *Investigation Report: Refinery Fire and Explosion and Fire, BP, Texas City, Texas, March 23, 2005*, 20 Mar. 2007.
5. The Baker Panel, *The Report of the BP U.S. Refineries Independent Safety Review Panel*, Washington, DC, 2007.
6. U.S. Environmental Protection Agency, *Chemical Accident Investigation Report: Tosco Avon Refinery, Martinez, California*, 1998.
7. U.S. Environmental Protection Agency, *Chemical Accident Investigation Report: Terra Industries, Inc., Nitrogen Fertilizer Facility Port Neal, Iowa*, 1995.
8. J. R. Taylor, Pre Incident Planning for Oil and Gas Plant, http://www.itsa.dk, 2014.

10 Diagnosis of Plant Disturbances

In a modern plant, disturbances will generally make their presence known by alarms, deviations in trend lines or high or low values on indicators. In many cases, these symptoms can be directly recognised and associated with a response.

If the circumstances giving rise to symptoms cannot be recognised, immediately the operator will need to diagnose the situation.

Some diagnostic responses are effectively 'rules', based on immediate recognition. If you see the bathwater running over, you shut off the taps.

This kind of knowledge is not immediately obvious as would be overfilling of the bathwater. On the other hand, the response was such that there would not be time for much reasoning. The operator had made an immediate association between frosting on the glass and excessive boiling up of urethane and between the frosting and potential for blockage and overpressuring (which was the fear driving the very rapid response).

CASE HISTORY 10.1 Instantaneous Visual Diagnosis

A supervisor was walking past a batch distillation unit, which was used to separate methanol and water from urethane. At the highest level of the plant was a 'cold trap', a glass spiral in a glass cylinder, with cold brine circulating, in order to condense methanol and prevent it from being released to the atmosphere.

The inner surface of the cold trap was covered with a white 'frost' (Figure 10.1). Within a fraction of a second the senior operator had started up the stairs. On arrival at the cold trap, he closed the brine supply valve and started to spray the cold tap with steam from a hose. Then, using a radio, he connected to the control room and told them to stop the heating to the distillation.

When I asked him later why he did not communicate with the control room right from the start, his answer was, 'There wasn't time for that'. He knew that the response time for cooling of the distillation kettle would be much longer than the time to reach blockage in the cold trap.

This kind of response could be made at this kind of speed only with some kind of ready-made recognition pattern. Since the event had not occurred before, the pattern must have been thought through or rehearsed earlier. Subsequent interview with the operator showed that he had many disturbances already experienced and visualised.

FIGURE 10.1 Urethane frosting in a methanol cold trap.

CASE HISTORY 10.2 Instantaneous Emergency Response

An operation manager noticed a small hole developing in a thionyl chloride drum. His 'correct' response should have been to sound an alarm and evacuate, especially since the ground was wet, and the release would generate sulphur dioxide and hydrogen chloride gases. Instead, he rolled the drum so that the hole was at the top, releasing almost no vapour. The reaction was 'instantaneous', i.e. within 1 or 2 seconds. (The drum was later placed in an 'emergency drum' which is oversized and made of tough plastic. The thionyl chloride drum was then set upright and emptied.)

Again in this case, the operations manager had a clear picture in his mind, of the drum, of its contents and of consequences. In this case the association definitely was pictorial; the manager had to visualise the level of the liquid in the drum.

These very quick associations between cause and effect and physical response strategies are described below. For an experienced operator, the repertoire of such responses may be very large.

The recall of such associations depends both on visualisation and on naming. Blowby is a phenomenon which can occur at the liquid outlet of a separator vessel, feed drum, distillation column, etc. It involves excessive flow of liquid, to the extent that there is none left in the vessel, and gas flows through the outlet. It is dangerous if the upstream pressure is higher than the downstream design pressure, because pipe or vessel rupture can occur.

It was found in experiments that once operators learned its name, they could quickly identify the potential and make control actions appropriately. They had more difficulty in explaining the physical basis for the blowby hazard (the flow resistance and, hence, the pressure drop across a level control valve are much less for a gas than for a liquid). This indicates that the knowledge in this case definitely is a learned association rather than a more extensive physical model. It contrasts with the two examples given above, where mental simulation of the situation was the determinant of the correct response.

The need for consistency and correctness of such immediate responses is such that a large fraction of operating manuals contain tables of such responses. The responses should be rehearsed by 'thinking through' the accidents and the responses, so that responses can be made quickly when needed. The quality of teaching for such responses can be important. Operators need to be taught visualisation of the disturbances. This needs more than just a single example shown by diagrams or by three-dimensional (3D) computer-aided design videos, because these can trap operators into a fixation on just one problem type. This happened for example in the Three Mile Island reactor meltdown, where operators were fixated on the problem of high water level in the pressuriser vessel, because this had been a major training issue. Operators need to be trained to create visualisations themselves, rather than to accept standard visualisations provided by trainers.

DIAGNOSTIC STRATEGIES

The descriptions in this section are based on minute-by-minute observations in a number of control rooms. The protocols for these studies are given in Ref. [1].

Immediate Associative Recognition

The ideal diagnostic strategy is to recognise the cause, and possibly also the sequence of events, and potential consequences based on experience, formal studies or previous mental simulation and rehearsal.

Sometimes, disturbances which are not so explicable occur. Examples are high-pressure drops in distillation columns and high temperatures in continuous reactors. The cause of these disturbances is often obscure.

Even when the immediate cause of a disturbance is reasonably straightforward, it is not generally possible to determine the condition of origin of the disturbance. Consider, for example, a high-level alarm on a gas/oil well fluid separator (two-phase separator) in an oil field system. There can be several causes for the high level:

1. New wells may have been brought on line, so that oil production is increased. The operator would normally be informed of this, so the expectation would be present, although the actual timing might not be known.
2. The flow regime in flow lines coming to the separator may have changed, so that a surge of liquid arrives. (This means effectively that oil quantities built up in the incoming lines are suddenly passed to the separator, usually because of flow instabilities as the quantity of oil in the oil/gas flow lines builds up. This is possible in some systems with a high gas-to-oil ratio.)
3. Bubble formation and a sudden level surge can occur because of a fall in the separator pressure. The pressure drop could occur because of gas-processing equipment coming on line downstream (depending on the pressure control scheme), or due to failure of the gas pressure regulation.
4. The level could have risen due to a blockage in the outlet flow line. This is a rare occurrence if the oil is light (contains a large fraction of light hydrocarbons, such as propane or butane) and would in most cases occur only gradually.
5. The level could have risen due to failure of the level control (the level control valve, the controller itself or the level sensor).
6. The cause is possibly to backflow or overflow from another separator.
7. Foaming is another reason (secondarily, because of a failure of injection of antifoam agent; only possible with some oils).
8. There is an increase in well fluid density, such as due to an increased water fraction (water cut) when level is measured by means of a pressure sensor (poor choice of instrument). This does not give a true increase in level, but it does give an increase in level indication.

Respond First, Then Explain

On seeing an alarm, the operators' first concern should be to determine how fast the temperature, pressure or level is rising, because this determines how much time is available before a trip level is reached. For example, in the level disturbance described above, if a level trip occurs, this may lead to a complete field shutdown, which would be very expensive (in some cases many millions of dollars). If only a short time is available, the operator would normally try to shut down some wells, to bring inflow and outflow into balance. On some plants, the operator would have a slug catcher vessel available, to which the inflow could be diverted, increasing the time available for diagnosis. In some plants, there is a large margin between the alarm level and the trip level, so that the separator itself can serve as a slug catcher. i.e. a vessel to take up large liquid flows.

Once it is determined that a good time is available or a time margin is established, it is possible to determine the possible cause.

Backwards Simulation

The strategy used systematically in some circumstances is akin to the forward mental simulation used to determine possible effect of a disturbance. The operator posits immediate causes, then causes of causes, then further back until a possible explanation for the disturbance is found. In the example above, an operator may suggest too high inflow as a cause and trace it back along the flow line to a change in the setting of the wellhead choke valve. It helps then to know who is in the field and the kinds of operations they are carrying out.

From experience in HAZOP (hazard and operability) studies, most oil plant operators know all of the causes, although they often have difficulty in recalling the last cause in the list above (experience from reviews of 12 HAZOP studies carried out in different companies). From experience in HAZOP studies, too, operators are not usually very good at organising this kind of search systematically so that all potential causes are found. (This is one of the reasons why it is necessary to have trained HAZOP leaders and systematic procedures.)

Assumption of Repetition of an Earlier Disturbance

The most common strategy for diagnosis of an event observed in the control room is to assume that the cause of the disturbance is the same as last time. This is so strong a tendency that it can be difficult to resist, even when other instruments show conflicting data (e.g. the case histories in the sections 'Piling on the Gas' and 'Reinterpretation as Instrument Error Normal in Chapter 15).

A similar assumption is that the cause of the disturbance is an anticipated cause such as, for example, expected wells coming on line in the example above. As before, such an assumption can quickly dominate thinking.

Before planning responses, such as reducing input flows, the operator should, as far as possible, check that his or her assumptions are correct. For example, if there is a flow sensor at the input, it is expected that the operator checks whether the inlet flow is greater than normal. Even better is to display a trend line for inlet flow, which can show the dynamics of the situation.

The value of confirmatory measurements is such that failure to check readily available confirming symptoms must be regarded as an error.

Brain Searching and Brainstorming

The fourth common diagnosis strategy in the control room is the more or less random proposal of causes, followed by a search for confirmatory evidence, first on control displays, and then, if necessary, by asking field operators to make checks.

In making the confirmation, reasoning or mental simulation is used, e.g.:

- 'Opening of wells gives an increase of flow in flow lines'
- 'Flow in the inlet manifold increases level'
- 'Flow into the separator increases level'
- 'Level controller cannot compensate increases level'

The modality of such mental simulations varies. For some operators the simulation is in terms of verbal expressions, usually much shorter than those above.

In applying these phrases, the operator obviously has to follow the topology of the plant, tracing the piping from equipment to equipment.

It has not been possible to determine whether the verbal phrases used to describe the simulation are the actual internal form used in thinking or are just an external expression, necessary for the operator to describe what is happening. What is certain is that this rather discretised chain of events–style description is used. This can be deduced from the details which are omitted and not even recognised.

A different modality is almost always shown by field operators, that of thinking and visualising flows. Operators can think, in this modality, of a surge of liquid coming along a pipe, pouring into the vessel and forming a wave of liquid which flows over internal weirs in the separator and causes the level sensor to rise and fall. It is characteristic that operators with field experience can visualise dynamics of disturbances, such as describing the effects of causes 2 and 3 in the cause list above. This kind of thinking is important, because it can reveal causes which are inaccessible (invisible) to operators with less experience and can lead to much better control.

A third modality has become available as a result of the 3D displays possible on modern workstation mimic diagrams. These show vessels in three dimensions often with cutaway wall, so that levels can be seen. Pipes are shown with two lines as walls, sometimes with liquid flowing, or with colour to show that the line is filled. Such displays encourage operators to think in terms of flows and accumulations. However, they necessarily suppress detail, and while helping to understand some problems, can serve to hide others. They can be dangerously misleading if the data displayed are erroneous, for example, as a result of instrument failure.

Systematic Causal Search

A fourth strategy for searching for causes is to search systematically for possible explanations. Each input line which could cause the high level by increased flow, for example, and each outlet line which could have reduced flow is searched in turn. The process of mental simulation is reserved. As far as could be determined, this could be done only by using the verbal simulation described above.

Each of these lines is followed backwards, searching for possible causes of the disturbance. Along the way, available symptoms are checked. Any symptoms which are incompatible with the simulated causes lead to that line of search being abandoned.

This kind of approach was only used occasionally by operators in the incidents studied and mostly by those who had been taught troubleshooting. In fact, systematic troubleshooting was only rarely necessary, unlike the first three approaches.

INSTRUMENT ERRORS

From study of accident reports it can be seen that instrument error, particularly a stuck instrument, can throw the diagnostic process either into complete disarray or onto the wrong track. This occurred in the Three Mile Island accident, the Milford Haven Refinery accident and the Texas City accident of 2005 (see Table 9.1).

This kind of problem should be much less prevalent in future. Firstly, instruments have become much more reliable, particularly since year 2000. Secondly, most new instrumentation in large plants has a considerable degree of self-diagnosis, using the HART protocol (a signalling method between control computers and instruments). Thirdly, there is a consistent trend away from the use of switches (temperature, level and pressure switches) to the use of transmitters, with a continuous measurement as output. (This is a requirement in several newer oil company standards.) Fourthly, modern DCSs allow process parameters as trend lines. A consistently flat line in a process, with no variations, is almost always an indication of failure or a parameter value which is out of range.

FIXATION

Fixation is a phenomenon which has been described by many authors, and which can be seen in accident reports, notably that of Three Mile Island. Once a diagnostic hypothesis has been found, and is corroborated by at least some corroboration, it is very hard to abandon this hypothesis, even when contradictory symptoms are found. This is even more the case if a complete team is engaged in searching for a course. Agreement is interpreted as confirmation. If 12 persons agree and the hypothesis becomes a fixed explanation, can the hypothesis possibly be wrong?

COMPLEX EQUIPMENT UNITS

The performance of some process equipments such as the separator discussed above can be understood with very simple rules, largely mass and energy conservation laws. ('If liquid flows in and does not flow out, the level rises'.) Some pieces of equipment are much more difficult to understand, and some are directly counterintuitive. Hydrocrackers, for example, generally have a quench gas flow to help reduce heating of the reactor bed, but under some conditions the quench gas can actually cause heating.

CASE HISTORY 10.3 A Simple Deviation with an Unexpected Consequence

In a large oil/gas-separation system, operation became increasingly difficult because the fraction of water in the oil increased with time. This tended to give water overflow into the oil fraction. This in itself was not a major problem—it required a slightly longer holdup period in the oil storage tanks to allow water to settle out, but the water had a drastic effect on the pumps transferring oil from the system. The pump seals failed much more frequently, about once every 2 months. (When the problem was reviewed, it was decided to replace the separator system with one of larger capacity.)

ERRORS IN DIAGNOSIS

Errors in diagnosis will in most cases lead to errors in response to disturbances.

Errors which can be observed from accident reports, and in some cases from observations in the control room, are as follows:

- Ignorance (lack of knowledge) of a physical effect, such as foaming.
- Errors in reasoning, e.g. failure of closing a valve, leading to increased flow.
- Overlooking a causal route: This may arise simply because a line is 'minor', because it does not have a flow normally or because it does not appear in the diagram the operator is using (whether mental or expressed on paper or on a display mimic).
- Failing to check for compatibility of a diagnosis with available symptoms.
- Ambiguity of the available symptoms (for example, in the level disturbance, it will be nearly impossible to discover causes if there are no flow sensors).
- Tunnel vision, restricting the range of the search, especially associated with the assumption that the cause is the repeat of an earlier cause, or the result of mislearning, that is, for example, teaching with focus on just one problem area, excluding others.
- 'Locking in' or fixation on a causal explanation because it makes sense for at least some of the symptoms: This can lead to complete teams ignoring conflicting symptoms and even direct indications. (This happened in the Three Mile Island accident of 1979.)

REDUCING ERROR POTENTIAL IN DIAGNOSIS

There are several techniques which can reduce error in the diagnostic process:

- Preanalysis of possible disturbances, e.g. using HAZOP study, and explicit tabulation of the disturbances, their symptoms, and potential consequences (see, e.g. Ref. [2])
- Advanced displays, such as mass balance derivation and display, hierarchical display, event sequence recording, sequential alarm playback [3,4]
- Ensuring that operators have adequate knowledge of the full range of causes

REFERENCES

1. J. R. Thomson, *High Integrity Systems and Safety Management in Hazardous Industries*, Oxford, England: Butterworth-Heineman, 2015.
2. J. R. Taylor, Does Quantitative Risk Analysis Help to Prevent Accidents?, To be published in *Int. Symp. Loss Prevention*, 2016.
3. ASM Consortium Effective Operator Display Design, Abnormal Situation Management Consortium, 2008 www.asmconsortium.com.
4. ASM Consortium Effective Alarm management Practices, Abnormal Situation Management Consortium, 2008 www.asmconsortium.com.

11 Errors in Decision Making and Planning

If a situation is unusual, it may require action which has never before been undertaken and which requires a previously unplanned action. This can go all the way from making small changes in procedures to extensive improvisations, which may involve significant engineering.

CASE HISTORY 11.1 A Difficult Decision Which Succeeded [1]

A unit producing light and heavy fuel oil was shut down for a period. The heavy fuel oil froze in the export pipeline to a ship-loading pier, 600 metres long. Heavy fuel oil is a bit like asphalt. Unless heated to about 60°C, it will not flow.

There was much discussion about how to bring the pipeline back into operation. The most extreme suggestion was to excavate a trench under it and burn oil in the trench. In the end the team assembled a few hundred welding transformers and three large diesel generators and heated the line electrically, by connecting power cables. The method worked.

Other less extreme examples of the need to make decisions and to plan are the following:

- The need to shut down piping which is leaking: Can the plant continue operating or does it need to be shut down?
- The decision on whether to shut down a plant where fluids are foaming and liquid is being transferred to a gas stream: Can operation continue, is it necessary to reduce the throughput or is a shutdown required?
- How do we remove a plug of plastic polymer which is stuck in a dump vessel?
- What do we do when all the product tanks are full and we no longer have storage capacity for the production?

In responding to situations like these, usually there is discussion about what the situation is, what the goal is and what should be done. Usually, it too involves some informal brainstorming.

Making the decision and selecting the approach requires knowledge. This includes knowledge about the plant, knowledge about the plant state, knowledge about engineering methods and knowledge about the underlying physics and chemistry of the plant processes. It can require knowledge of the physics of accidents.

Errors in this process are typically as follows:

- A method which does not work is selected. This is not always a real error; sometimes you have to try before you really understand the problem.
- The necessary preconditions for the new plan are not understood, or not known. Insufficient attention is given to what could go wrong.
- The possible side effects or downstream effects of the plan are not known.
- A plan which has a high level of risk is created.

CASE HISTORY 11.2 A Decision That Failed

An example of insufficient attention of preconditions for safety is given in an incident documented by the U.S. Chemical Safety Board [2].

A pinhole leak was discovered in a crude unit on the inside of the top elbow of the naphtha piping, near where it was attached to the fractionator. The operators responded immediately, closing valves to isolate the line. Operation was continued. It proved difficult to isolate the line and drain it, a situation which indicated that some valves were 'passing', that is, leaking internally. The leak reoccurred, and the valves were tightened.

A plan was made to replace the line without shutting down the plant. A work permit was issued which allowed the maintenance team to drain and remove the pipe. The draining was unsuccessful. The supervisor directed workers to make two cuts in the piping by using a cold cutting method. The cuts immediately began leaking naphtha. As the line was being drained, naphtha was suddenly released from the open end of the piping. The naphtha ignited and quickly engulfed the tower structure and personnel. Four were killed.

It was later found that a valve at the discharge end of the pipe had been passing, and pressurised naphtha had leaked backwards into the pipe.

Apart from the frightening continuation of production, which involved a strong element of risk taking, possible hidden or latent hazards were overlooked. The pipe could be back pressured; the possibility that downstream valves could pressurise the pipe were overlooked or ignored. The preconditions for safe operations were not checked.

CASE HISTORY 11.3 A Decision without Considering Distal Consequences

A company produced a silicate-based window sealant which included zinc powder in the mix. One batch was mixed wrongly and began to generate hydrogen. The supervisor told the employees to send the material to the chemical waste disposal plant but to leave off the lid until the last moment when the drum was placed on a railcar.

When the drum was received at the disposal plant, it was lifted onto a trolley by a labourer. The drum exploded, and the lid cut into the man's chest. He survived but even after recovery was too disabled to work.

In this case, the supervisor did not think far enough about the effects of his plan.

CASE HISTORY 11.4 Improvisation without Analysis

Figure 11.1 shows a bent propane drainpipe. It was caused when a forklift truck was used to support a pushrod (wooden spar) used to clear stuck product from a vertical blowdown drum. The idea was not bad in itself; it meant at least that an operator did not have to stand underneath the discharge hole cover, pushing the rod himself and waiting for material to fall. The hazard of the rising truck fork interfering with the pipe was overlooked, however. The lift of the truck hit a propane pipe and bent it. Luckily the pipe only bent; the connections at the ends of the pipe could have broken, with a potential for a large fire.

RISK TAKING AND RISK BLINDNESS

Taking risks is a fundamental and necessary part of some aspects of process plant operation. The kind and the level of risk vary, but these need to be kept under control.

FIGURE 11.1 Pipe bent by a forklift truck during an improvised dump tank–emptying procedure. Improvisations of this kind sometimes have to be made. It is important, though, before trying something new or of doubtful safety, that the hazards are identified the risks are assessed and precautions are taken.

CASE HISTORY 11.5 Risk Taking

Georgie was an experienced rigger, with a long experience and a remarkable sense of balance. With help from his younger colleagues and a good crane operator/driver, he could manoeuvre a 50-ton pressure vessel into place through structural steel framing with as little as 1-inch clearance.

Georgie was sitting on a 6-inch wide beam at about 20 metres height as a new tank was hoisted into place! He could adjust the lift alignment with a minimum of effort, with just a slight push from his fingertips.

A slight wind moved the vessel an almost imperceptible distance. One of the vessel nozzles caught on the lower steel structure. The vessel tipped, cutting Georgie's legs at the thigh. He was dead by the time he was brought down from the high steel.

Riggers are a race unto themselves. Many rigger teams do not mix with other construction workers. They are often quite fearless. They are always confident in their own abilities—the job would be impossible without this.

In a modern, well-run construction site, the procedure used for alignment in the case history above would probably be forbidden, but Georgie had carried out the same kind of operation perhaps a hundred times. His confidence in his own expertise and his experience made him blind to risks. The accident was not the result of his own actions except that of putting himself in danger. The direct cause of the accident was an external force, the wind and the ever-present risk of loads catching on surrounding steel.

I have tried to determine the cause of such hazardous behaviours in interviews. One of the causes which becomes quite clear from these is lack of imagination coupled with lack of direct experience of accidents. Experienced operators, fitters and maintenance artisans, especially the most skilled, are susceptible to this. A typical comment in a session of job safety assessment is, 'I have worked in this job for over 40 years and have never seen that happen, so I don't think it is possible'.

A successful strategy I have found for avoiding problems of this kind involves the following:

1. To require job safety analyses to be carried out for potentially hazardous tasks
2. To use a large collection of accident photographs to illustrate the safety analysis; there is nothing like a good accident picture to convince a doubting operator

The form of risk blindness shown in the case histories can arise in explicit decision making during planning or in the instinctive performance of tasks which are routine.

The main problems with this type of risk blindness are the following:

- Inability to judge the level of risk
- Habituation, so that the hazards gradually become invisible

The ability to judge the level of risk, i.e. the probability of the risk, depends on experience of actual accidents and near misses. Accident reports posted on notice boards can help to widen experience. Even better are job safety reviews. These can well be the topic of toolbox talks.

RISK HABITUATION

Habituation involves performing a potentially hazardous task repeatedly with no adverse consequences until one day all goes wrong.

CASE HISTORY 11.6 Risk Habituation and Extreme Risk Taking

Opening filters in pipelines is often necessary in order to prevent blockage. The procedure involves closing the upstream and downstream valves, then opening the vent valve on the filter so that any pressure that can arise inside is released. Pressure can arise, especially if either the upstream or the downstream valve is leaking (passing) or simply because of the residual pressure of operation.

In the actual accident in a paint factory, the vent valve gradually became blocked by paint residue, and the filter remained pressurised. When the maintenance worker opened the filter (by unscrewing four screws), the filter cover was not immediately blown off, because the gasket stuck. When he tried to remove the gasket with a blow of a hammer, the cover blew off, breaking his jaw.

Really good operators become cautious with time; they expect trouble. 'Never walk under a crane, never stand in line with a blind flange and never trust a ladder until checked' becomes a mantra. For others, habituation is a truly dangerous phenomenon.

There are other forms of risk blindness. One is hazard blindness, in which the person simply does not know that something is dangerous.

CASE HISTORY 11.7 Hazard Blindness

Two fitters were asked to empty a tank of phosphoric acid and to remove deposits blocking the outlet nozzle. Emptying the tank would take a long time, and production would be halted during the entire job. They decided to loosen some of the bolts on the manhole cover to speed up the process (80 bolts in all, all difficult to free).

When about 30% of the bolts were loosened, the bolts, being overloaded, and the manhole flange 'unzipped'. The manhole cover was blown off by the weight of the acid, and the 60°C warm acid flowed out. One fitter was killed; the other was very seriously injured.

Lack of knowledge of physics, along with insufficient experience to make physical judgements, is one of the main causes of the more serious accidents in process plants. A description of some of the more serious phenomena is given in Ref. [3].

One of the most prolific sources of data on risk blindness comes from the operation of machines. Numbers of cases and frequencies of incidents were collected and are presented in Ref. [4]. There are large numbers of cases of operators reaching into or climbing into machines without ensuring first that the machine is isolated and locked out.

DELIBERATE RISK TAKING

Some risk taking is an essential part of plant operations. Not all problem situations can be foreseen, and not all problem situations can be recovered without taking some risk.

In the bromination reactor incident described in Case History 15.11, operators and the supervisor knew that starting the agitator on a reactor in which a large quantity of bromine had accumulated was dangerous, with a possibility of reactor runaway. Not knowing any other way to respond, they decided to start the agitator anyway.

CASE HISTORY 11.8 Considered Risk Taking

A blower providing air to a sulphur burner developed a heavy vibration, detected by all vibration sensors installed on it. The operators shut down the plant and stopped the blower and waited for the unit manager and operations engineers to arrive.

The unit managers were faced with two choices, that of restarting the blower to see whether the cause of the vibration had gone away and that of dismantling the blower and inspecting the impeller. Dismantling and inspection would take several days and probably require a specialist to be flown in from Europe. Several days of production would be lost. It should be recorded that the blower was no small piece of equipment; the impeller was 2 1/2 metres in diameter and weighed over 400 kilogram and normally rotated at 3000 rotations/minute.

The managers decided to start the blower 'carefully', with the operator's finger hovering over the shutdown button. Before reaching full speed, one of the impeller blades, weighing 65 kilogram, broke off, was ejected into the discharge ducting, and broke a hole when it hit the first ducting elbow. Sulphur dioxide flowing back from the sulphur burner poured out like a waterfall, creating a large plume of toxic gas.

The impeller continued to rotate, because of the momentum, even though the blower was shut down quickly. The now unbalanced impeller caused enormous force and vibration, ripping out eight 2 1/2-inch bolts from the shroud support concrete and ripping out or shearing 24 bolts from the main impeller bearing. The friction of the axle rubbing on the damaged shroud ignited lubricating oil, starting a 12 metres diameter pool fire.

On the good side, the fire was put out within 10 minutes, and the system was rebuilt and restarted within 12 weeks. No one was injured. The investigation showed that the problem started at a small imperfection in the impeller metal, which developed into a fatigue crack. The blade tore off when the fatigue crack became large enough.

It is known from interviews that the plant managers never envisioned the extent of the damage which could be caused. They took a risk with limited knowledge (which is almost always the case in risk taking). The balance of risks was not in their favour.

Whether the actual restart can be called an error or not is moot. Operators and operations managers have to take risks sometimes, when unusual situations arise. If the gamble pays off, they are regarded for their expertise. If it does not, their action is condemned as error. The only true errors in this kind of situation are in not performing an adequate risk analysis, preferably before the fact, and in not seeking all possible information before going ahead.

RISK TAKING AND RISK BLINDNESS IN MAINTENANCE

There are cases in which safety regulations and rules are deliberately broken. The reasons are varied; laziness, desire to finish a job quickly, high spirits and exuberance and the conviction that safety regulations are unnecessary nonsense. Examples of deliberate risk taking are the following:

- Man riding on transport belts
- Reaching into machinery to clear blockages
- Opening equipment without going through proper isolating, inerting, and venting sequence
- Entering confined spaces without checking
- Not performing routine scheduled checks

Many people are to some extent guilty of negligent risk taking. Take, for example, exceeding the speed limit by a small amount or not using a car safety belt. It is correct to speak of guilt in this context, because the laws and regulations are clear. If an accident occurs, and the breach of laws or regulations is found to have contributed, punishment can be expected.

Negligent risk taking becomes particularly serious, firstly, when the consequence becomes severe; secondly, when the probability of accidents becomes high and thirdly, when the resulting injuries are to others. Operating a process plant beyond its safety limits and failing to test the safety system are typical serious forms of negligence in plant operation management. Completing fraudulent records, or claiming that tests have been carried out or procedures are followed, exacerbates the culpability and in many countries is a criminal act.

Reviewing a number of cases of deliberate risk taking, several features are apparent:

- There is considerable pressure to meet production targets or to meet a restart schedule. The pressure can be related to pressure from senior management or be just derived from professional pride in doing a job effectively.
- Those responsible do not believe that the risk is high. They underestimate either the consequences or the probability.

- A frequent factor is that persons are ignorant of the true status of safety systems. They assume that they are protected by 'defence in depth'. For example, they may assume that operators will be able to stop accidents, even if alarm systems are out of action. They also rely on the conservatism built into safety assessments.
- There is no appreciation of just how wrong things could go.
- The person is unable to imagine the hazard.
- There is often ignorance of just what the safety regulations are.

Judging the probability of deliberate risk taking is difficult, because it varies so widely. Even in a well-run plant, with a management which is risk aware and which enforces safety regulations, sometimes there are individuals who take risks.

CASE HISTORY 11.9 Convenience and Efficiency before Safety

During a major maintenance turnround on an oil and gas plant, several people were found working inside a vessel which had not been gas tested. The reason (excuse) they gave was, 'The gas testing technician was overloaded, and we could not afford to wait the 2 hours that it usually took for gas testing to be completed. In any case, full gas freeing and ventilation had been carried out'. In the investigation it was found that bypassing of gas testing was becoming a standard practice.

It is common in some plants for those responsible for approving permits to work (PTWs) to do so without visiting the work site. This to a large extent invalidates the whole PTW process. In many cases it can be seen that the PTW system *cannot* work properly, because it is not adequately staffed. This is, from observation, the most frequent example of deliberate risk taking in otherwise well-run process plants.

Preventing deliberate risk taking is one of the main objectives of 'safety culture' development. The author has been able to observe the development of what could be called a safety culture over periods of several tens of years. At the start, managements were concerned about safety and followed all the good safety practices which they understood. (Risk consultants are not invited to work in plants where there is no safety culture.) Over a number of years, practices gradually improved, including understanding at all levels. The key is that the most effective steps were as follows:

- Managers should be visible and should show good safety behaviour and safety interest
- Development of clear risk management guidelines
- Development of good practices in HAZOP and job safety analyses, required for *every* plant and *every* task
- Development of good and effective PTW systems and extensive training in its use
- Gradual increasing understanding in the company workforce itself and among the company's contractors

For high-integrity organisations, there is nevertheless some risk arising from individuals prepared to bypass the rules, just like there are car drivers prepared to ignore speed limits. Sometimes, deliberate risk taking becomes directly fraudulent.

NECESSARY RISK TAKING

Deliberate risk taking is not just a question of error. Many legitimate tasks require a degree of risk taking. The objective is to ensure that the degree of risk is not excessive.

CASE HISTORY 11.10 A Measured Decision

During preparations for a heavy lift, a soft spot was found in the asphalt-covered ground, so that one of the crane supports began to sink. Reinforcement was obviously necessary. Two possibilities were found. The first was to place timber balks across the area and cover these with steel plates. The second was to excavate the location and make an engineered hardstanding. The second solution was obviously safer, since the actual state of the ground could not be known. However, it would delay the project by several days and cost several million dollars as a result. After discussion, the second, safer solution was taken, even though calculations indicated that the first would be safe enough. The reason given for the decision was that the first solution involved too many unknown factors.

Such well–thought out risk taking is frequently necessary. A major group of such activities involves tasks which depend on the weather. A sudden change in weather conditions can affect erection of plant, loading and unloading.

Risk-analysis techniques have been developed so that routine risk taking can be well controlled. Risk-taking problems arise when unusual, unexpected or completely novel situations arise, or when there is insufficient knowledge to allow an accurate prediction to be made. Project and production managers then need to take decisions under conditions of uncertainty and, in some cases, ignorance of the true conditions.

CASE HISTORY 11.11 A Difficult Evacuation Decision

A leak occurred on a flare header (large pipe leading to a flare stack where gas from relief valves is burned). The gas contained hydrogen sulphide, which was, and is, highly toxic. Gas detection worked, and alarms caused workers in the workshop that were immediately downwind to take shelter indoors. The question then was whether to evacuate the persons or to continue to let them shelter in place. The evacuation would be risky, involving them moving through the gas plume. Letting them stay in the building would also be risky, because of gas gradually seeping into the building. Shutting down the plant would have made the hazard worse, because shutdowns in this kind of plant produce short-lived increases in pressure and additional relief through safety valves. In the event, two firemen and two fitters wearing breathing apparatus very quickly fitted a clamp to stop the leak temporarily until the flare could be shut down and a permanent repair could be made.

Success in such decision taking depends on knowledge, ability to think through a task to its end, imagination to determine what can go wrong and a degree of caution. It also requires an ability to analyse the physics of the situation.

CASE HISTORY 11.12 A Weird Decision

A rather small, but amusing, example of decision taking (mis)behaviour arose during testing of a chemical waste incinerator. The incinerator was to be fed with a small dose of methylene chloride (very volatile liquid which is mostly chlorine, but in organic and relatively nonhazardous form) in order to determine the degree to which dioxin formation could be avoided. A pump was to supply methylene chloride through a small injection nozzle. The temporary piping was difficult to arrange, since a hose connection was needed but the diaphragm pump available was overdimensioned. No smaller pump available could be used because methylene chloride is a 'magic solvent'. It is able to destroy most seals so that only the Teflon-lined pump would work.

The work was carried out, unusually, by the plant manager and the commissioning coordinator (myself) late in the evening on the day before the tests were to begin. The coordinator insisted that the injection piping should be tested using water before testing with methylene chloride. In the first test, the hose coupling failed, being forced off the injection pipe by pressure pulses from the pump. The clamps were tightened, but the hose blew off again. Grooves were cut into the pipe, but this did not work either. As the evening wore on, many different kinds of hose clamps were tried. All failed.

Finally, the plant manager said: 'This isn't working with water; let's try with the real stuff'.

The surprise was that as far as I could see, the plant manager meant this seriously.

(The next day, a proper welded and flanged piping arrangement was made.)

INSTITUTIONALISED RISK TAKING

Before commencing this section of the book, I should state that I have been fortunate enough to work with many companies which can be described, in the terms of La Porte and Consolini, as high-integrity organisations, where risk taking is reduced to a minimum and where risks are taken only after careful analysis. Inappropriate risk taking would be further from their thoughts than would be stealing from the petty cash. These companies consistently strive to define best safety practice, then work hard to achieve it. They continually strive to update their understanding of best practice and to learn from around the world of the best safety (and operational) methods.

And then there are the others. In accident reports it is quite common to find the following:

- Plants were undermanned.
- Operators had inadequate training.
- Known deficiencies were not remediated, or in some cases were not even scheduled for remediation.
- Safety assessments were of poor quality.
- Operating procedures were out of date or just plain wrong, and updating was not carried out or in many cases not even scheduled.

Such problems can arise due to organisational ignorance. Companies as a whole, particularly small companies, may not know what is required for safety operation.

Other problems arise from risk blindness. The company knows, for example, that there is a risk of fire on storage tanks and that the fires can extend to enormous sizes in a boilover. However, they are unable to see that 'this could happen here' or that 'the risk is too high; we must do something about it'. In discussions I found that senior managers often resisted ideas about large accidents. Only after having seen videos of actual accidents on plants similar to their own could they conceive that 'yes, it could happen here'.

There are several solutions to this kind of problem:

- Third-party loss-prevention or risk audits, carried out by loss-prevention engineers from outside the organisation: These persons have the authority of the collected experience of their companies and also the power given by the need to sign an audit report before insurance terms are agreed.
- Formal risk analyses carried out competently (after 40 years of working with risk analysis, I still doubt the effectiveness of this approach except in organisations which have embraced risk assessment and are used to making use of the results; photographs and videos especially work better).
- Making use of lessons learned from accidents around the world, especially those with good photographs.

ASSESSING RISK-TAKING HAZARDS

It is virtually impossible to predict the frequency of risk-taking behaviour. If a company operates with a poor safety culture, with little emphasis on safety, or if a company as a whole is relatively ignorant of good safety practices, then it seems likely that risk-taking behaviour will occur. Risk calculation for such companies seems rather pointless unless something can be done to change the behaviour. By studying the frequency of accidents, for example, from Accidental Release Information

Program and Risk Management Plan data, and by correlating these with information from major accident reports, it is possible to see that actual risk in a risk-taking organisation is at least 10 times that in a well-functioning high-integrity organisation. Table 9.1 above gives a list of some major hazards and indicates those where risk-taking behaviour played a role, according to the accident reports.

Assessing the tendency for risk-taking behaviour is more important for an organisation which has otherwise high-integrity behaviour. From studying accident reports it is likely to arise when there are specific difficult tasks and when there are work overloads. Working in remote areas, or areas with difficult access, is also a situation which encourages risk taking.

Another institutional cause of risk taking arises when safety-related tasks, such as testing and inspection, are simply insufficiently staffed, or for which insufficient time is allocated. A common example is testing of emergency shutoff valves. If testing requires plant shutdown, it is common that any delays in shutdown tasks during main overhaul result in incomplete testing.

In all, it seems better to plan in such a way that there is no incentive to improper risk taking and which ensures that any such cases which do arise are detected quickly.

REDUCING THE POTENTIAL FOR RISK-TAKING ERRORS

The job safety analysis, task risk analysis and management of change procedures should have as one of their main purposes the goal of identifying adverse preconditions, possible hazards and possible unwanted side effects in decision making and plans (see Chapter 21).

REFERENCES

1. U.S. Chemical Safety Board, *Investigation Report: Tosco Avon Refinery Martinez, California*, 1998.
2. U.S. Chemical Safety Board, *Case Study: Oil Refinery Fire and Explosion, Giant Industries*, 2004.
3. J. R. Taylor, Industrial Accident Physics, http://www.itsa.dk, 2015.
4. J. R. Taylor, *Risk Analysis for Industrial Machinery*, ITSA, 2002.

12 Infractions and Violations

Not all plant operations go according to the rules. Shortcuts may be taken; standards may be lowered; unauthorised improvisations may be carried out to get the job done, to reduce workload or to reduce effort and sometimes rules may be deliberately broken for personal gain or because of disagreements.

These deviations from good operating practice may vary from minor infractions to violations requiring immediate dismissal. Leaving a hose connected after completion of a job, rather than coiling and storing it properly, is abandonment of a minor infraction. Smoking in a refinery is a serious violation of rules (and good common sense) and usually results in dismissal.

Identifying such practices and taking them into account in safety analyses is difficult for working plant and impossible at the design stage. Identification and risk assessment is nevertheless important, firstly, because such infractions and violations can add significantly to risk and, secondly, because in many cases it is relatively easy to do something about them.

HOUSEKEEPING AND EVERYDAY HAZARDS

Poor housekeeping is important because it inculcates a feeling that no one cares and contributes to general poor performance. It also creates tripping and dropped object hazards. In an emergency it can slow access for rescue, release control and firefighting.

Figure 12.1 shows an example of poor housekeeping. This example is ironic. The plant safety officer stated during the audit that he did not feel that the standard was too bad. About 30 minutes later he stepped on a small length of scaffold pipe, slipped and hurt his back.

UNAUTHORISED IMPROVISATIONS

In most safety departments in large process plants, there is a 'black museum' of improvised tools and methods. Figure 12.2 shows an improvised shutoff for an air hose. The problem with this kind of improvisation is that if the wire fails, the hose can whip around. Many injuries have been caused in this way.

Figure 12.3 shows an improvised sulphur pipe heater made from a steam hose and two heavy bolts. The main problem with this 'design' is that if the bolts are dislodged, the pipe will whip around, with a potential for severe injury to anyone caught nearby.

Figure 12.4 shows an explosion ignition safe power supply receptacle and plug. It has been illicitly 'adapted' to allow ordinary household power cable and connection adapters, creating a serious ignition source.

FIGURE 12.1 A relatively mild example of poor housekeeping.

FIGURE 12.2 An improvised block valve consisting of a bent hose and a length of wire.

FIGURE 12.3 Steam hose used to heat a sulphur pipe.

UNAUTHORISED EQUIPMENT

Use of unauthorised equipment is common, especially among contractors, unless very active steps are taken to prevent it. Audits are needed to find this kind of problem. Use of ordinary tools rather than Ex safe tools, use of sparking tools and use of inadequate cranes or bulldozers and excavators as cranes have frequently led to accidents.

MACHO BEHAVIOUR

Some workers take pride in getting things done and will take personal risks to ensure that the work is fast and efficient. Figure 12.5 shows a laboratory technician registering drums of chemical waste. It is nearly impossible to do this job from the side of the truck. A gantry should have been provided to allow registration to be done safely. But in the meantime, the technician is not going to wait or complain.

Within 2 weeks of this photograph being taken, a worker on another site close by put his foot through the lid of a drum, suffering cuts, chemicals burns and very nasty bruising as a result.

FIGURE 12.4 Ordinary household cable connected into an explosion-safe power supply, found in a plant making extensive use of flammable solvent.

Similar problems can be seen with drivers climbing on the top of road tankers and balancing on the cylindrical top of the tank in order to fit hoses into the filling hatches.

In some countries the macho attitude is endemic. In one Eastern country I pointed out that plant construction workers were wearing felt boots. If they trod on scrap metal, they could suffer serious injury. It was explained to me that the workers needed the felt boots to ensure good footing and balance when they were walking along pipes.

FIGURE 12.5 Registering drums of chemical waste by walking on them.

CASE HISTORY 12.1 Unauthorised Bypassing

A catalyst module factory was built with safety in mind. Many robots and fully automatic production machines were constructed, so that only two persons were needed in the entire production line. They checked the product for quality.

The production was designed for maximum safety. It was completely fenced, and the doors were interlocked to the production. Opening a door or gate stopped the production conveyors and robots. It was even fitted so that the machinery could be tested without opening the doors.

The safety was nevertheless a nuisance for one machine fitter. The safety circuits were easily bypassed by using two ladders to help climb over the 2 1/2 metres fence.

Cases like this can result in single-person injuries, sometimes fatalities, but the can-do attitude can be much worse.

During maintenance it is usual to remove power from pumps and other electrical devices. A typical measure is to pull the circuit breakers away from their bus bar power connections and then to lock the electrical cabinets. One electrician, finding a cabinet locked, and without the correct key (in the pocket of the person carrying out maintenance), found his work delayed. The problem was easily overcome though. He unscrewed the cabinet door.

This kind of behaviour can cause major accidents. Short circuits and accidental contacts in circuitry which should have been locked out have caused pumps to start, valves to open and conveyors to start. Valves opening unintendedly are particularly liable to cause accidents if the reason for lockout is that the piping has been partly dismantled.

Bypassing of interlocks is necessary in some situations, such as when emergency shutdown circuits are to be tested. Modern practice is to require all bypasses of interlocks to be approved by the plant manager and to be registered in the control computer. The best systems send an e-mail to the plant manager if an interlock is bypassed for too long a period.

CASE HISTORY 12.2 Ignoring Safety Precautions

A crane used to take household waste from hoppers and feed it to waste incinerator boilers was to be made fully automatic. Safety interlocks were bypassed to allow software testing. An error in the software caused the crane to go out of control and its grab to smash through the control room window, narrowly missing the programmers.

HORSEPLAY

Some employees like a joke and some like to joke around. This kind of behaviour is best kept for the playing field, however, not in a process plant.

A tragic example is young people who found an old 'empty' tank to be a convenient place to relax and enjoy conversation. It was situated in an open area, so there was really nothing wrong with their behaviour until they decided to see what was in the tank. One of them lit a scrap of paper and dropped it into the tank. Vapour in the tank exploded.

ILLICIT BEHAVIOUR

Smoking is universally banned except inside designated smoking rooms in all refineries, gas plants and petrochemical plants because of the danger of igniting flammable vapour. Smoking is also forbidden in many chemical plants. Cigarette lighters are also forbidden. Near truck loading and unloading stations, however, it is common to see cigarette ends on the ground.

DELIBERATE FRAUD

Occasionally simple cheating or fraud occurs with the objective of saving money or saving effort.

CASE HISTORY 12.3 Fraud in Plant Inspection

In a nuclear power plant, radiographic plates used in checking for weld defects had been overexposed. A test piece, with different thicknesses of metal, is used to indicate proper exposure on the plates. The plate had been shaded with pencil, to give the appearance of proper exposure.

CASE HISTORY 12.4 Another Inspection Fraud

On another nuclear plant, two pumps were to be radiographed to check the quality of welding. To save time, one pump was radiographed and two copies of the same radiography plate submitted for quality control.

CASE HISTORY 12.5 Fraud in Material Supply

In audits of materials delivered for construction projects, the author found that up to 30% of the steel ordered as high-grade material was actually ordinary carbon steel, even though the plates and piping were stamped as alloy steel. In one oil-processing plant, over 1000 bolts and nuts were found to have been the wrong alloy type.

Supply of nonoriginal parts seems to have been less of a problem for process plants than, for example, in aircraft maintenance. However, some cases were found, including valves which were refurbished in used equipment, with flange faces built up with weld material and then remachined.

VANDALISM AND SABOTAGE

These rather unpleasant subjects might be thought to be rare, and for most of my personal experience, these are rare in process plants. According to the 5-year incident histories given in the U.S. Risk Management Plan database, some 2% of refinery incidents with off-site consequences are caused by vandalism or sabotage. The records are made by the refineries themselves so the information needs to be interpreted in that light.

CASE HISTORY 12.6 Sabotage

A strike occurred at a chemical plant, and an audit was required to ensure that the plant had been shut down safely. It was found that the plant was in safe condition, but lubricating oil supply to pump bearings and seal fluid connections on pumps had been disconnected. Any attempt to start would have ruined the machinery.

Figure 12.6 shows the controller for a fire detector. It was sabotaged with a screwdriver cut across the printed circuit board because the detector was giving continued intense alarms and could not be shut off.

PREVENTING VIOLATIONS

The most usual way of preventing serious breaches of safety rules is dismissal. In many of the places where data were collected for this book, vandalism and sabotage

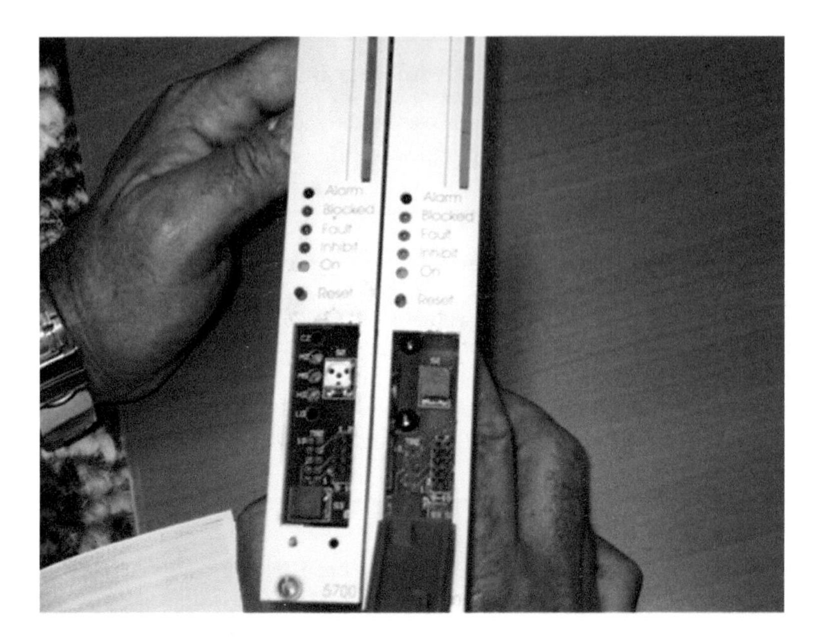

FIGURE 12.6 Fire detector controller sabotaged by a cut in the printed circuit.

are unheard of. The jobs are just too precious to risk losing them, and for most persons the thought of sabotage would never occur.

Macho behaviour and unapproved improvisation can best be prevented by inculcating good safety culture. There are known techniques for this (see Chapter 21) but it does take time and effort and is completely dependent on management leadership, both from the top and from middle management.

Preventing fraud requires a degree of policing, which is usually carried out by quality control. The use of positive material identification, using X-ray emission spectrographs, has greatly reduced the problem of substandard materials.

13 Management Error

Management error is almost always a feature in major accidents as evidenced by root cause analyses in accident reports.

By far the most common types of management errors are omissions. In some cases managers do not know what is required of them. In other cases, they doubt the value of many safety-oriented procedures and implement them only reluctantly. In still further cases managers do not wish to implement safety procedures.

CASE HISTORY 13.1 Refusal to Follow Industry Best Practice

During a safety audit it was learned that an oil production platform did not operate a permit to work (PTW) system. Such systems are a crucial measure in preventing accidents during maintenance or unusual operations. The operation manager refused to implement such a system and instead implemented what were called 'safe procedures'. Contractors were required to read these and to sign them after reading. The reason for this system was that there was little room on the platform to allow a PTW office team. The discussions became quite acrimonious, and the safety officer in the audit resigned.

About 6 years later, a small team of contractors visited the platform in order to perform various maintenance tasks. One task was to replace a level switch used as an alarm on a compressor knockout drum (used to prevent liquid from entering the compressor and destroying it). The team shut the inlet valve to the knockout drum but did not drain the drum, which contained a mixture of liquid ethane, propane and butane. Also, they did not stop the compressor or shut the suction piping to the compressor.

When the level switch was removed, the compressor sucked air into the drum. The air mixed with propane vapour, and when the mixture reached the compressor, it ignited. The gas burned back towards the drum and built up to a true detonation. A 1.2 metres wide section was blown from the drum wall. The explosion was so violent that the flame was blown out, allowing the propane and butane vapour from the remaining liquid to escape and spread through the platform. The gas found an ignition source and a vapour cloud explosion occurred. Eleven workers on the platform were killed, including the maintenance team. The damage resulted in losses of about $1.2 billion.

PTW systems are intended to prevent just this kind of accident and, if operated properly, ensure that an experienced person inspects the area, including proper shutdown and isolation, before the equipment is opened. Systems which rely on self-checking and on reading instructions are much less safe and depend on many assumptions, not least the assumption that the maintenance team has someone who can read.

ERRORS OF OMISSION IN PLANT MANAGEMENT

The following is a list of omissions in safety documentations and procedures which are characteristic of older plant management styles:

- Lack of operating procedures, out-of-date operating procedures and lack of periodic review of procedures
- Lack of cautions and warnings in procedures
- Lack of a matrix or manual of permitted operations
- Lack of defined operating limits for the plant; even when there is such a table, there is often a lack of defined heating and cooling rates for the plant
- Lack of HAZOP studies or out-of-date HAZOP studies; HAZOP studies archived without being read
- Lack of a management of change (MOC) procedure or lack of any safety assessment as part of this procedure
- Lack of a PTW procedure
- Lack of the requirement that no PTW may be signed without visiting the site where the work will be carried out
- Lack of a lockout–tag-out procedure
- Lack of emergency procedures or lack of location-specific preincident plans
- Lack of inspection programme
- Long backlogs in carrying out inspections and preventive maintenance
- Omissions in testing programme
- Lack of training programme
- No requirement for job safety analysis or task risk analysis
- Lack of follow-up of agreed actions from HAZOP, hazard identification, environmental impact identification and quantitative risk assessment studies and safety audits

INADEQUACIES IN SAFETY MANAGEMENT PROCEDURES

Complete omission of necessary safety actions is straightforward to identify. More difficult

- Less than adequate (LTA) overview of design
- LTA training
- LTA staffing provisions for maintenance, testing and PTW
- Poor quality of safety analyses
- Superficial MOC assessments
- Lack of management understanding of safety analyses
- Insufficient and incomplete plant inspections and integrity assessments
- LTA training for emergency response, especially management training
- Excessive trust in consultants, with limited buildup of in-house expertise

VIOLATIONS

Some omissions are so extreme that they are regarded as breaches of the law. Many examples can be seen in U.S. Occupational Safety and Health Administration (OSHA) audit reports, which are published in many U.S. states. Examples can be seen in British practice, especially from court proceedings. Some examples are the following:

- Start-up with incomplete testing
- Excessive testing backlog of critical safety devices, such as pressure safety valves, blowout preventers, emergency shutdown systems and fire-protection systems
- Out-of-date drawings
- Out-of-date procedures
- Ignoring audit reports
- Inadequate training
- Operating with plant not certified as fit for purpose
- Arbitrary downsizing of operating and maintenance activities
- Improvisation without safety assessment

MISTRUST OF SAFETY ANALYSIS AND SAFETY MANAGEMENT

Over a period since 1974, with the publication of the first major public risk analysis (the Reactor Safety Study, [1]), the field has moved from a research area to a specialist academic area and then to a central activity in authority regulation of industrial safety. In many oil, gas and chemical industry plants, risk analysis has become a working tool. In the plants in which I spent most of the last 15 years, risk-analysis regulations were rolled out in 2001, and by now any operation supervisor is expected to be able to perform a task risk assessment and to use a risk matrix in order to determine applicable safety measures within his or her area of authority.

Nevertheless, an institutional aspect of risk taking is that many companies (as well as individuals) have very little understanding of or respect for risk analysis and modern safety engineering practices. (The names of some of the companies are most easily read in the accident investigation reports of the U.S. Chemical Safety Board.) The mistrust, lack of respect or just plain lack of care is reflected, for example, through the following:

- Disregard for the importance of safety
- Disregard for safety documentation, regarding it as useless paper
- Attitude that safety will be adequate 'just so long as the design meets the standards'
- Attitude that safety activities are a luxury which can be minimised when profits are down

In the early 1970s, when I started my career in safety engineering, safety departments were very understaffed compared with today's practice. The post of plant safety officer was often a retirement post, where persons could be parked as an alternative to terminating their employment. This can still be seen in some companies.

In some companies too, there is an attitude where production, efficiency and minimum overhead are objectives in their own right, and all 'unnecessary' overheads which do not contribute to production should be minimised or eliminated. This is the opposite of good safety culture and is generally punished when the plant breaks down or blows up. Unfortunately, it is generally not the guilty who suffer the punishment.

Modern safety management is well staffed and very bureaucratic, with health, safety and environment (HSE) plans, control of major accident hazards reports, environmental impact assessments, hazard and effects register, HAZOP and safety integrity level reports, inspections plans, audits, job safety analyses and PTW forms. At their best these documents are useful working tools. If they are made as cheaply as possible and only to satisfy a regulatory requirement, it is quite likely that they will be valueless, confirming prejudice against safety practices and can be more dangerous than helpful.

CASE HISTORY 13.2 Unminimised Regret

I once had a conversation with a plant manager who later became a good friend. His plant had a chlorine production unit, and a rather unusual accident resulted in a release of about 80 kilograms of chlorine. The cloud passed over a kindergarten, and, although no one was hurt, the incident naturally caused concern, not least among the parents. A grassroots organisation made serious protests and tried to have the plant closed. There was strong political influence to do just that, but there was also strong influence to keep the plant open in order to preserve employment. The compromise was to make an in-depth risk analysis, and the company undertook to implement the recommendations. The result was what must be the most secure chlorine plant in history, with all the storage tanks and pumps in concrete bunkers and with a big inherently safe scrubber to absorb any possible chlorine release. There were many other improvements.

During the studies, the atmosphere was acrimonious; the employees and the management did not like the presence of risk analysts in the plant and would have been quite happy to exclude the whole analysis team. In the end, a working compromise was reached, and the plant allowed to continue operating 'until another possible site could be found'.

About a year later there was a major release of hexane in another part of the plant and an unconfined vapour cloud explosion which more or less destroyed the plant itself. The area was converted to a car park.

Later, the plant manager joined the committee convened to write a green paper on the implementation of the Seveso Directive. An industrial representative with personal experience on large accidents was an ideal contributor. His wistful comment to the explosion—'I did not realise it at the time, but I wish we had asked your team to analyse the other half of the plant'.

CASE HISTORY 13.3 Pressure to Minimise Safety Effort

Another example of lack of appreciation of hazard analysis occurred recently. The occasion was a session to analyse a reciprocating gas compressor system. A very similar system had been analysed earlier, and the project manager exerted considerable pressure for us simply to copy the earlier analysis and review just the differences between the original and the revised system. Considering that HAZOP study meetings with the usual participation cost the company about $10,000 per day, and the work had been done once before, the pressure was perhaps understandable. However, the original analysis had been done in a 'quick-and-dirty' style, with only three nodes assessed and with only two critical accident scenarios identified. The second analysis was done with 12 nodes and was thorough, even though it took just half a day. Twelve very significant hazard scenarios were identified, which were not covered in the first HAZOP study. One of these was really serious, the possibility of liquid flow into the compressor, giving rupture of the cylinder head or the piston lead and usually leading to a large fire or explosion. This is a well-known phenomenon, with many accidents on record. Because of the emphasis on doing the job quickly and cheaply, the first HAZOP study was worse than useless; it could have led directly to an accident.

The vendor representative who took part in the HAZOP study confirmed that these accident types did occur occasionally, that they were the most frequent accident types and that the consequences would be catastrophic. He also confirmed that safeguards against these were available as options!

The example shows how insidious the transition to a poor safety ethic can be. The project manager has experience that HAZOP studies do not contribute much to safety (at least the ones he had experienced). He had a duty to limit expenditure and meet budgets. He was unaware of the hazards.

It would be easy to dismiss this manager as unsuitable for the job, not knowing enough to build a safe plant. This is not a constructive approach. Safety culture cannot be built by dismissing managers (although in isolated cases it has helped). Safety is built by good examples and positive experiences. Very often in a good HAZOP study, for example, there is an 'aha' moment, where even the most extreme sceptic realises that improvements are needed.

Fortunately, over the years there has been an increasing emphasis on safety. Reviewing the U.S. Risk Management Plan records [2] (with about 22,000 plants reporting their 5-year accident records for accidents with off-site consequences), it is possible to see that a large percentage (I estimated 50%) had achieved at least the intention of good and thorough safety culture, and many of the rest had at least a focus on several important aspects of safety. Unfortunately, accident reports reveal that there are still many companies with poor safety management practices.

I myself have been fortunate enough to work mostly with companies who have a good safety culture in place and wished only to improve. This is probably a self-fulfilling situation, since only companies with a concern for safety are likely to hire risk analysts and consulting safety engineers. The companies with poor safety culture have more generally hired me to investigate accidents.

Over the 45 years in which I have been able to observe, there has been an enormous improvement in the norms for safety practices. As just a small example, in the late 1960s, riggers would rely on experience, judgement and balance to avoid falling. In many plants now, any failure to use a safety harness and to tie off safety lines properly is ground for dismissal. This applies for all work more than 1.8 m off the ground unless there are safety nets.

LACK OF KNOWLEDGE OF PROCESS SAFETY

One of the issues which has come to be recognised over the last 15 years or so is that there is a large difference between safety at work and process safety. Work safety is largely concerned with preventing everyday accidents, such as falls, electrocutions and traffic accidents. It is a very important task and can turn a potentially hazardous workplace, such as a refinery, into one which is just as safe as an insurance office. Process safety is concerned with making plants that do not blow up, burn down or release clouds of toxic gas. The techniques used are very different from those of safety at work, largely those of HAZOP analysis, quantitative risk analysis and integrity assessment. The best companies now have separate departments or sections for these activities. The process safety or technical risk management departments are generally staffed with safety engineers, a profession which did not truly exist 20 years ago.

CASE HISTORY 13.4 Multiple Managerial Failures

An example of the problem is illustrated by the Macondo deep-water oil well blowout of 2010 [3]. This accident resulted from a combination of some judgement errors, at least one design error and several errors in lack of maintenance. Most of the problems had been analysed in depth earlier and extensive risk analysis reports made, but the lessons from these had not been taken to heart. By contrast, managements from both of the companies involved had assembled on the drilling rig just 2 days earlier to award prizes for good work safety performance. They were not uninterested in safety. They just did not understand that good safety at work, protecting workers from everyday accidents, does not mean that you have good process safety.

The lack of awareness of the need for process safety and for technical risk management was understandable until the beginning of this century (i.e. 2000). Most senior managers have a background in production, and work safety has been a part of their daily lives, in particular dealing with accidents and reporting work accident statistics. The lack of attention to technical safety could then be regarded as lack of knowledge. Today, so much emphasis has been placed on this issue of process safety, especially in the Cullen report on the Piper Alpha accident [4] and in the Baker report on the Texas City 2005 accident, that lack of awareness of these issues should be regarded as culpable ignorance.

DANGEROUS DECISIONS

Most of the management errors above are errors of omission. Sometimes though, managements contribute actively in causing accidents.

CASE HISTORY 13.5 A Dangerous Decision [5]

A light gas oil side-cut line on a crude distillation unit was found in an inspection to have significant thinning corrosion. Part of the pipe was replaced, although the inspectors recommended complete replacement. Later a small leak was found, but before it could be shut down, the leak caught fire. The fire was put out, but firemen were then ordered to remove lagging so that the source of the leak could be found and clamped. The leak got worse, then the line ruptured. Firemen may have damaged the pipe when trying to remove insulation cladding. The large inventory of high-temperature gas oil was released and formed a huge vapour cloud. The managers ordered shutdown of the unit, but by this time it was too late. After about 2 minutes the cloud ignited, causing an extremely large flash fire. Of the 19 persons close by, attempting to deal with the leak, none were killed. Treated for smoke-related effects were 15,000. The cause of the thinning was high-temperature sulphidation corrosion, which affected the whole pipe.

The U.S. Chemical Safety Board investigated the accident and found the following:

- The company did not effectively implement internal recommendations to help prevent pipe failures due to sulphidation corrosion.
- The company failed to perform internally recommended 100% component inspections.
- The refinery's turnaround planning group rejected the recommendations to 100% component inspect or replace the portion of the side-cut piping that ultimately failed.
- Some personnel participating in the insulation removal process while the four-side-cut line was leaking were uncomfortable with the safety of this activity because of potential exposure to the flammable process fluid. Some individuals even recommended that the crude unit be shut down, but they left the final decision to the management personnel present. No one formally invoked their stop-work authority.
- Decision making encouraged continued operation of a unit, despite hazardous leaks. There had been an earlier example of continued operation, despite there being a fire.
- There was reluctance among employees to use their stop-work authority. Recent safety culture surveys performed at the refinery indicate that employees had become less willing to use their stop-work authority.

There were 24 such observations in all. In 2013, the company pleaded no contest to 6 charges in connection with the fire and agreed to pay $2 million in fines and restitution.

There are many accidents like this in the lessons learned database. In retrospect we may well ask, 'What were the managers thinking?'

In such situations the managers were certainly not thinking, 'Here is a situation which can blow up my plant, ruin my year-end profits, possibly kill people and wreck my career'.

Some insight into thinking can be obtained from earlier history. On April 10, 1989, an explosion and fire occurred in a cracking column at the refinery. A total of eight workers and firefighters were injured. Three workers suffered second- and third-degree burns. As a result, in September 1989 the OSHA fined the company $877,000 for 'willfully failing to provide protective equipment for employees'. Employees had 'repeatedly requested' protective equipment since the early 1980s but the company had refused despite more than 70 fires in the plant since 1984 [6].

One of the main influences in this kind of operational culture is the urgency of production. When a refinery crude distillation unit is shut down, it can cost many millions of dollars per day. Any shutdown, even for a small correction, is expensive, because, although the units can be shut down quickly, it takes at least a day to get back into operation. Also, major repairs such as replacing a side-cut pipe require draining, purging, steaming out and ventilating the affected area, or even the entire unit, and this takes several days.

Losses of tens of millions of dollars can readily be seen in refinery production statistics and key performance indices. One of the managers' dominating goals in their work is to achieve as high key performance indicator (KPI) values as possible (if they are ambitious) or to ensure at least that these do not fall below the norm (for most managers). The motivation is even stronger if the refinery is running with a narrow profit margin, or is losing money, as can be the case for older and less efficient refineries.

There is also an influence from habituation, of becoming used to accidents and regarding them as a natural fact of life. This is usually the result of tradition, with attitudes formed at the beginning of the life of the refinery, often 50 years ago, in some cases 100 years ago. The management has become used to accidents, fires and even fatalities. It has become a cost of doing business. By contrast, in newer refineries, any release of hydrocarbons, and especially any fire, is rare. If it does occur, there will be an independent investigation and a major effort to learn lessons. There can be career penalties for managers if they are found to have contributed to accidents, even when the companies operate a no-blame culture.

A further part of the story is the belief that 'we can manage the plant, and we know better than anyone else how to deal with incidents'. The intensity of this belief is revealed by the defences made during OSHA fine appeals. This kind of view is based on experience. The managers are for the most part not in denial or not completely so. Their experience, often built up over 30 years or more, is that there may have been problems, but they have defeated them. 'That is why we have a fire service. We can handle the accidents'. This is often said with a certain pride. 'I have been in the industry for 40 years and have never had a major problem'. To which the reply should be, 'I will come back in a thousand years and see if that is still true, because the target is a major accident frequency of less than one in a thousand per year'. Even this target would result in one major accident per year in a country like the United States, and the usual risk-analysis target is less than one major accident per hundred thousand years per plant, with frequencies above this requiring efforts for risk reduction and as low as reasonably practicable (ALARP) analysis.

The background to this kind of thinking is a kind of lack of knowledge, or ignorance. Managers often do not know what major accidents have occurred and often do not know the details, even if they know of a particular accident. In one emergency response training course at a company with very good safety culture, I showed a video of a large jet fire which occurred when contractors cut into the wrong pipeline. The first comment was, 'We had a mistake like that last month. Fortunately, the line was empty. I didn't know that the fires could be that big'. In the actual case the fire was so large that firefighting was impossible due to heat radiation. The emergency plan for the company being trained described how the firefighters would cool down the entire area with water sprays from fire water monitors, but the execution of the plan was physically impossible. The firefighters would never have been able to approach close enough.

Very often, people who have devoted a lot of time, effort and money into safety cannot believe that their efforts can be insufficient or ineffective until it is demonstrated to them in reality. If they are fortunate, this will be at the time of an emergency exercise.

It is not surprising that in some companies there is little knowledge of major accidents. Only rarely is training provided for senior managers. The training needs to be specific for their kind of plant; otherwise, there is an immediate reaction—that cannot happen here—which robs the training of any effect. The training, therefore, must be very focussed and needs to be supported by risk analysis or audit, so that the managers can be shown just where and how the accident could occur, accompanied especially by video presentation of accidents which have occurred.

A comment from an operation manager is telling. After a large emergency training exercise, in which he had to play the role of incident commander, he courageously stated, 'I cannot do my job in this kind of emergency. I have not been trained for this kind of work'. His statement was meant as a request for help, from a person who obviously was highly competent at his job and wished only to become more competent.

Problems have occurred at the highest command level in many of the emergency exercises in which I have taken part. Just because an operation manager is highly educated and experienced, we should not assume that he or she does not need training.

PATCHING UP AND MAKING DO

The use of clamps to seal leaks in the response described in Case History 13.5 is also telling. Clamps are often used to stop small pinhole leaks caused by pitting corrosion, especially in low-pressure pipes and water pipes, and are very useful for this purpose, especially in emergencies. Clamps do not work well when the cause of the leak is metal thinning corrosion. Indeed, clamps can crush the pipe in such situations if the pipe walls are very thin (see Figure 13.1 to understand the makeup

FIGURE 13.1 A clamp used to stop a leak through a crack; leaking gas is piped to a safe place.

of a clamp). Welding on patches, too, is only a very temporary repair. You know that you are in trouble when there are three clamps, or when you start needing to patch the patches. Even if they work for a short period, it is only a matter of time before another leak occurs.

Clamps are even worse when used to stop leaks from cracks. The clamp may stop the leak, but it cannot stop the crack growth. A growing crack will at some stage lead to a rupture and, on hydrocarbon or chemical equipment, to a catastrophe. Again, there is a knowledge gap, which will tend to make the risks of the situation less obvious.

Nevertheless, managers who think of clamping are not thinking of catastrophe. They are thinking of a drip or pinhole leak. Managers do not normally worry about pinhole leaks; the maintenance staff can look after such things. The managers are not thinking of a large jet of hydrocarbon, and even if they do consider them, they usually think of a pool fire developing, not a large jet spray causing a 200 metres wide cloud and a massive explosion.

A final question needs to be asked. The managers in Case History 13.5 clearly made a mistake—the mistake of being caught out by the physics and the corrosion. But how often had pinhole leaks occurred before? How often had clamps been used and proved to be a successful way of maintaining operations?

If we are to overcome problems like this, there is a need for hazard awareness training at the highest management level, and it needs to be relevant, specific and practical. Also, it is no use just describing problems; the trainers need to be able to provide practical solutions.

A side problem of the case above is the question of how it was reported, especially the framing of the report. There is a big difference between a report which says, 'We have a pinhole leak and a drip of kerosene', and a report which says, 'The thinned area of pipe x has finally failed. It is only a small leak just yet, but it is growing'.

DOWNSIZING AND OUTSOURCING

One type of institutional risk taking arises from what is in principle a good practice. Companies in a large group compare performance. The objective is to find which processes and which practices are most efficient, how to achieve best-in-class operations and generally to achieve best profitability. One of the issues often addressed is that of whether the organisation is overstaffed.

The one thing that is difficult in such comparisons is to take safety into account. There is at present no way that the value of safe operating practices can be quantified in the same way as productivity. Risk-analysis approaches as used at present in the process industries generally ignore the issue of human error and even ignore the benefit of human intervention. In many jurisdictions it is even forbidden by authorities to take credit for good safety management practices.

This is not to say that safety comparisons are not made between companies. Inclusion of people from other plants into process safety management audit teams is a standard practice in many companies and is very effective in improving safety. The problem arises because such intercomparisons are not part of the profitability assessment. As a result, staffing levels, training practices and provision of high-quality emergency equipment tend to be de-emphasised or in some cases regarded as a clear area for saving.

One of the practices which has been encouraged in this drive for efficiency and productivity has been that of downsizing and outsourcing maintenance and even in some cases operations. Again, there is in principle nothing wrong with this practice provided that it is done properly. It can even improve maintenance practices if the contractor or 'partner' selected is really competent. However, having seen this in operation in many companies, it is possible to confirm that in no case was the impact on safety considered when outsourcing of in-plant maintenance activities was introduced. The clear loss of experience in many such cases, especially with termination of the older and higher paid employees, has often caused concern about how the companies could survive such loss of experience. In a significant fraction, they did not.

By contrast, I have been able to observe the effect of extreme attention to safety in large turnround maintenance projects, in some cases setting records for safety.

Downsizing was a very significant development in the United States in the 1980s and served to improve productivity. At the same time, there was a steady increase in major accidents in process plants in the United States, reaching a peak in 1989, with 10 major explosions, among which was the Pasadena polyethylene plant explosion Case History 13.6. In many of the accidents, including the Houston explosion, outsourcing of maintenance was found to be *a* major contributor to accident causation.

CASE HISTORY 13.6 Catastrophic Maintenance [7]

One polyethylene production process involves mixing liquid ethylene under pressure with propane or isobutane to dilute the ethylene and prevent a runaway polymerisation reaction. Catalyst is added in the form of microscopic crystals of aluminium oxide. The mixture is pumped around a loop. The polyethylene is generated as 'fluff' or 'popcorn' in the mixture.

It is necessary to bleed a part of the circulating mixture continuously and to separate off the diluent so that it can be returned to the process. This is done in a so-called settling leg, and the product is released from the bottom of the leg.

The arrangement is prone to plugging, as the polyethylene polymerisation process continues. In the actual case, three out of six settling legs were plugged, and a maintenance company was engaged.

Under the company's written procedures for this maintenance function, which was usually performed by a contractor, company operations personnel were required to prepare the product settling leg for the maintenance procedure by isolating it from the main reactor loop before turning it over to the maintenance contractor to clear the blockage. On Sunday a contractor crew began work to unplug the three settling legs. According to witnesses, all three legs were prepared by a company operator and were ready for maintenance, with the reactor isolation valve in the closed position. The air hoses, which are used to rotate the valve, were disconnected. Number 1 leg was disassembled and unplugged without incident. At approximately 8:00 AM on Monday morning, work began on number 4 leg, the second of the three plugged legs.

The contractor crew partially disassembled the leg and managed to remove part of the blockage from one section of the leg. Part of the blockage, however, remained lodged in the pipe 12 to 18 inches below the reactor isolation valve. At noon, the contractors went to lunch. Upon their return, they resumed work on number 4 leg. Witnesses then report that a contractor employee was sent to the reactor control room to ask a Phillips operator for assistance. A short time later, the initial release occurred.

Five individuals reported actually observing the vapour release from the disassembled settling leg. Because of the high operating pressure, the reactor dumped approximately 99% of its contents (85,200 pounds of flammable gases) in a matter of seconds. A huge unconfined vapour cloud formed almost instantly and moved rapidly downwind through the plant.

In the investigation the pneumatic control hoses for the valve were found to have been replaced the wrong way round. The valve opened when it should close, and vice versa. The valve opened; isobutane and ethylene were released and ignited, causing a massive explosion, killing 23 and injuring 314.

There were many findings from the investigation, including the fact that there was no blind flange isolation, as was required by corporate procedures, which were not actually followed on the plant.

The reason for the hose reconnection was never fully clarified, but the reason for connection the wrong way round was easy to understand. The connections were identical.

Since July 1972; OSHA had conducted 44 inspections in operations of the contractor company at various locations; 7 of these inspections were in response to a fatality or catastrophe, and another 17 were in response to employee complaints of unsafe or unhealthful working conditions. The 44 inspections resulted in citations for 62 violations (including 19 serious) and $12,760 in penalties. Of these inspections, 2 were conducted at the facility where the explosion took place.

OPTIMUM SAFETY? PROBLEMS WITH THE ALARP PRINCIPLE

Relying on cost–benefit analyses to determine the 'correct' level of safety is not always appropriate. In many cases, in times of poor economic climate, process plants are not profitable. This has been true for long periods in the oil-refining business, for example. Plants are kept running at a loss in such periods in order to avoid the shutting down of a going concern and in the expectation that business will get better. In such a climate, investments for safety may be very difficult to justify on a cost–benefit basis.

One way of stabilising this kind of risk taking is to establish norms and standards for safety. The American Petroleum Institute Recommended Practice 750 for process safety management was an early start to this and was followed, after the U.S. Senate hearing into the Phillips, Houston, accident, by the requirements for process safety management in the Code of Federal Regulations Title 29 CFR 19.40. Another set of good practices was introduced as a consequence of recommendations in the Cullen report on the Piper Alpha accident [4].

Having been involved in safety planning for several major maintenance campaigns (which are necessarily outsourced in most cases, because of the number of persons involved), I hope that the following lessons for outsourcing of maintenance have been learned:

- Make sure that there are good standards, method statements and procedures for the work before entering into a contract.
- Make sure that good quality and safety standards are part of the contract, with bonuses for safety and quality.
- Make sure that whatever the completion bonus is, it is nullified if there are serious safety or quality infractions (which should be defined).
- Develop a good PTW system, and staff it with company staff, at least in the area authority positions.
- Set up a good audit system, and staff it with experienced maintenance managers.

PITY THE POOR PLANT MANAGER

One of the serious problems which can arise in business life is that business does not go so well. If the company is actually losing money, things become really bad, and safety improvements come very low down on the priority list. Generally, the highest priority is the payroll (although some cases of staff working months without pay are known). Even higher priority may be given to purchasing raw materials if credit limits have run out.

You will rarely be involved in a HAZOP analysis, let alone human error analysis, under such circumstances unless the regulating authorities (labour inspectorate or environmental authorities) insist on it. When this does happen good low-cost safety solutions are needed for the problems found. In the one case where I was involved in such circumstances, it was surprising how much safety could be obtained at no cost at all.

THE PARACHUTE MANAGER

A problem which has appeared repeatedly in accident reports over the last 40 years, and particularly after the 'lean, effective management' revolution of the 1980s is that of managers who move into a position for a period of just 2 years and who have the objective of 'turning round' a plant to achieve efficiency and profit. This in itself is not a bad thing; profits through efficiency are good for nearly everybody and everything, including safety. What is wrong is not the goal but in many cases the way of achieving it.

The easiest way to achieve profit in a process plant is to cut staff, particularly those with a long employment history and large pension entitlement. This approach is much easier than trying to make the plant more technically efficient or to derive better throughput through improved performance or debottlenecking. Such improvements take much skill and generally take longer than 2 years.

A further way to reduce costs is to reduce maintenance, which not only saves the cost of maintenance but also reduces plant downtime and allows more continuous production.

Of course, reducing staffing and reducing maintenance is not without risk. In the end, someone must pay. However, by the time this happens, the parachute manager has moved on, possibly moved up, as a result of his or her 'success' in increasing profit.

In some cases, the manager may believe he or she is doing a good job. In others, the behaviour can be so extreme that it can only be described as sociopathic. The manager knows that there will be a price to pay, but simply does not care. Examples observed of this (from personal experience) are managers illegally disposing of toxic waste, operating plants with major sections of safety systems inoperative, postponing critical repairs, falsifying inspection records and dismissing injured persons in order to maintain a good safety record.

CASE HISTORY 13.7 Hubris [8]

An example of where a manager appears to really have believed that reductions were properly made was for an oil-production platform in South America. A company executive is quoted:

The company has established new global benchmarks for the generation of exceptional shareholder wealth through an aggressive and innovative cost cutting programme on its production facility. Conventional constraints have been successfully challenged and replaced with new paradigms appropriate to the globalised corporate market place. Through an integrated network of facilitated workshops the project successfully rejected the established constricting negative influences of prescriptive engineering, onerous quality requirements and outdated concepts of inspection and client control. Elimination of these unnecessary straightjackets has empowered the project's suppliers and contractors to propose highly economical solutions with the win-win bonus of enhanced profitability margins for themselves. The project shows the shape of things to come in the unregulated global market of the 21st century.

Hopefully, the last sentence is untrue, because the platform had two serious explosions about a year later, with the loss of 11 lives, and sank 5 days later, with a capital loss of US$0.5 billion and considerable delays in oil field development. The accident commission for the incident [9] described the causes of the sinking as follows:

- A design error which allowed hydrocarbons to enter the emergency drain tank directly from the production header rather than from the production caisson
- Delay in starting the emergency drain tank pump for 1 hour
- The failure of actuators to close ventilation dampers, allowing water to flood flotation compartments
- Two seawater pumps being under repair, without emergency measures in place in case of emergency
- Inadequate contingency plans and inadequate training for dealing with emergency ballast and stability control situations

From reviewing accident reports, aggressive cost cutting appears to have cost shareholders worldwide over $40 billion in asset losses and business interruption over a period of about 20 years, just taking the largest 100 accidents into account.

LEADERSHIP

In reviewing the analyses and audits of 130 plants, belonging to over 90 companies which formed a background to this study, the one thing that sticks out foremost concerning the safety practices is that plant safety is determined by senior management knowledge of good safety practices and their leadership in applying the principles. The plant will be a high-integrity plant if

- Managers visit the operating plant frequently and are visible;
- Managers wear proper personal protective equipment at all times when in the plant and obey safe working rules;
- Managers establish clear safety rules, then provide the workforce to support the rules;

- Managers discipline employees who break rules;
- Managers do not allow unauthorised improvisations and new ways of working without a job safety assessment;
- Managers provide proper safety training;
- Managers ensure that needed documentation, drawings and procedures are up to date and readable and
- Managers follow up and ensure mitigation or prevention or repetition of all incidents and near misses and of all hazards found in studies and workshops.

CASE HISTORY 13.8 A Success Story

A chemical waste plant had a major hazard accident once per year over a period of 7 years. There were many demands from the local community for the plant to be closed. As a result, the management were forced by the authorities to carry out a thorough risk assessment, which included a full safety audit.

Following this, a period of 30 years has followed with no further major accidents, although there have been two near misses.

The risk analysis itself made some difference in that several weak points in the design were eliminated. All modifications on the plant have since been subject to risk analysis. However, these changes alone can explain only a small fraction of the improvement. At a follow-up after 25 years, the practices established after the risk analysis were either still in place or had been improved on. The management had changed, but only by natural progression; the original plant manager still worked for the company as a consultant.

The changes can be traced directly to a change in knowledge and practices of the management. There was never any doubt about their good intentions or their determination to run a safe plant. The analysis provided them with the tools needed to understand their risks and to do something about them. In comparison with this, the recommendations made on the basis of the risk analysis provided a small contribution.

In more recent years I have had the opportunity to work for a group of companies with a very high level of integrity imposed (literally) from the top. In spite of an international workforce, it has been possible to see over a period of some 12 years a continuing improvement in safety knowledge and in safety practices. True leadership works. This does not mean that performance is perfect, and there are continuing problems with management of contractors, since each new project seems to generate its own problems. Importantly though, the management system has sufficient checks and balances to ensure that the majority of problems are caught. In the one area where performance was inadequate, with several accidents, the companies in the group all made major (and expensive) efforts to eliminate as far as possible all risks associated with the problem.

SAFETY CULTURE

Safety culture has been mentioned several times in this book, but what is it? A simple answer is that good safety culture is a condition in which everyone in a plant follows good safety practices instinctively, without question, and does not deviate from good safety practice.

This does not mean that there will not be accidents. Mistakes or slips can be made. It does mean that if the safety regulations and practices are complete, there will not be any high-risk situation.

Good safety culture requires leadership from the top, a serious effort to convince middle management and training to enable all the working staff of the measures which are part of the culture. It also requires knowledgeable and competent HSE professionals.

One of the important methods of achieving good safety culture is that people are first *trained* in good safety practices and then *assessed* on their performance. KPIs which in some way measure safety have been an important part in development of good safety culture.

Good safety culture means much more today than it did even 20 years ago. Expectations have risen. As just one example, in many plants today, no maintenance is carried out without a method statement and a task risk assessment. Twenty-five years ago, the term *task risk assessment* did not even exist.

As a result of developments in HSE techniques, safety has improved for many companies. As an example of the effect of this, 20 years ago lost time accident rates of less than 1 per million hours were considered good. Today companies are reaching rates of 1 per 30 million hours in many cases.

One of the main new steps in safety culture which has become a cornerstone of recent advances is the realisation that good performance in preventing accidents at work does not necessarily mean safety against catastrophes. Different techniques and efforts are required. As a result, process safety, including HAZOP studies and other similar safety reviews, have become an important aspect of safety culture and all operators and maintenance technicians are expected to be able to participate.

Moving from a poor safety culture to a good one is difficult, and examples will be described in the following section.

DRIFT INTO FAILURE

Dekker, in an insightful book [10], describes the drift into failure occurring when an organisation is pressed, especially economically.

In the period covered here it was possible to observe the internal workings of just a few stressed companies.

Companies with poor safety culture and performance are often those which were established long ago and have never drifted out of a state of poor failure performance. Fatalities in process plants were regarded as a normal fact of life, in the same way as we now regard road traffic accident fatalities. The performance of some of these companies is deteriorating and appears to be getting worse as the plants get older. The U.S. Chemical Safety Board has reported several accidents arising due to corrosion in pieces of equipment which were up to 50 years old.

One company which was possible to observe closely was an old one, built up in a heroic location and in a heroic age. Employees had a tough, can-do mentality, which meant that sometimes the (un)safety practices were frightening. As an example, in one case a contractor was given a PTW which included grinding. There were insufficient power sockets at the site, so cables were stretched through an area where flammable liquids were being drained to the ground. The cable was made of coupled lengths, joined by ordinary household connectors, creating an enormous fire risk. This case was not exceptional. The fatal accident rate for the company was 12 per 10^8 hours, which compares with an oil industry average around the world of 4.4 per 10^8 hours and best performers' average of 0.8 per 10^8 years.

Motivated by the chief executive officer, the company struggled to achieve better performance, which was eventually achieved. The main obstacles were middle management and supervisors, who 'knew how to get the job done'.

A surprising company was one that was really stressed, to the extent that it could not always afford to buy raw materials. An in-depth risk assessment and safety audit was required by the authorities, and during the course of that, the operations were found to have a very high level of safety. The environmental performance, though, was very poor, and considerable investment would be required both for cleanup and for improved operational facilities. One week after the safety assessment, the company was closed because of poor environmental performance. The company was too small to be able to bear the cost of a full environmental protection infrastructure.

This example is important in showing that poor safety is not inevitable in spite of economic or operational stress in a company.

A third example is that of an oil-production company with a falling level of oil production. With falling revenue, the company was forced to cut costs. Also, unfortunately, the amount of work needed in an ageing oil field does not go down. If anything, it increases. Tasks were outsourced, employees were offered retirement packages and others were encouraged to seek new jobs. Mastery of office politics became a major career skill. The most technically competent sought contracts elsewhere. Over a period of 2 years the company staff reduced by 30%.

Obviously, something had to give, but safety remained high for a period. Over the next 10 years, however, the accident rate gradually increased, including process plant accidents.

Drift into failure is definitely a significant effect. Many older oil installations are being pressed economically now, from large new oil refineries and petrochemical plants built closer to the source of the oil and, at the time of writing, from low oil prices. Many chemical plants in the Western world too are pressed, especially from low-cost producers in China and India. It will be important in future to remember that safety is a good investment. Accidents on average cost about 2% of turnover, and even when the losses are insured, insurance companies always collect *their* losses over a period of years.

HIGH-INTEGRITY ORGANISATIONS

Many companies in the oil, gas and chemical industries can be classified as high-integrity organisations. Such organisations have both good safety culture and good safety practices [11]. The companies have low accident rates. Typical fatal accident rates were quoted above as about 4.4 per 10^8 working hours. A high-integrity company can have fatal accident rates one-third of this, with the spectrum of causes dominated by transport accidents. This can be compared with 21 per 10^8 hours for the construction industry in the United Kingdom and 20 per 10^8 hours in the service industries.

Major accidents cannot be compared so easily, but the frequency of accidents with off-site consequences can be determined for the United States, from data collected under the U.S. Environmental Protection Agency's (EPA) Risk Management Plan, notably data from the 5-year accident histories. Over 5 years in the 1990s there were 364 incidents with off-site consequences reported for 68 refineries over roughly 5 years (some companies overreported), giving an average frequency of 1.07 per refinery year. There were 36 fatalities and 284 worker injuries. This average, though, covers a wide range of performance. Some companies had no incidents at all; some had as many as 12. There is obviously a wide range of safety performance. Reporting was audited by the U.S. EPA, so most of the difference is unlikely to be in reporting practice.

High integrity is achieved by hard work. A list of company guidelines from a typical high-integrity company gives some impression of the extent of the work:

- Health and safety policy
- Health and safety objectives
- Occupational health and safety manual
- Communication, consultation and employee participation
- Health and safety documentation and archiving
- Health and safety audit
- Nonconformities review and corrective and preventive actions
- Incident reporting and investigation
- Lessons learned implementation
- Hazard and risk assessment, inherent safety review, HAZID, HAZOP study, 3D model reviews, fire and explosion safety assessments, area classifications and human factors
- Health and safety management system, threat and barrier identification and barrier integrity management
- Safety critical systems' performance standards and verification
- Emergency preparedness planning and emergency exercises
- Health and safety performance measurement and monitoring
- Health and safety training and certification
- Job safety guidelines and regulations (26 regulations)
- PTW system
- Job site inspections
- Task risk assessment

- Equipment inspection programmes
- Plant mechanical integrity audit
- Inspection and quality control for new installation and repairs
- Third-party safety audits
- Transport remote monitoring and company transport safety services

The whole system involves a lot of paper and is bureaucratic, but the bureaucracies are efficient (they have to be). It can be seen from the list that barriers have been inserted into practices for all known accident types. Gaps in the defences are usually closed whenever they are identified. This can take time, though, because it often involves writing guidelines, developing a training programme, actual training and implementation. Current typical gaps in defences are in the quality of the risk analyses made and some gaps in communicating the systems to contractors.

All this work does not eliminate accidents entirely. There are still accidents, although in the companies reviewed the accident rates were lower than the averages for the industries involved and for the industry overall, as described above.

The pattern of accidents is also different for high-integrity companies. Except for road traffic accidents, most of the accidents which do occur have 'strange' causes, often involving hitherto unheard-of accident mechanisms, new chemistry or new forms of corrosion. As a result of this, at the time of writing there is a movement towards much more extensive use of lessons learned, especially from other companies.

ASSESSMENT OF MANAGEMENT ERROR?

There have been several attempts to develop approaches which allow the quality of management to be assessed and incorporated into risk assessments, for example [12,13]. Some of the attempts have little credibility, because they attempt to calculate an overall 'management factor' by which overall risk values as calculated can be multiplied in order to derive a 'true value risk'.

More successful are methods which attempt to determine risk levels for specific types of accidents, such as fires due to hot work during maintenance. The effect of management omissions or limited intervention in these cases can be assessed by determining the effectiveness of PTW systems, and the degree to which workers are trained for hot work, for example [14,15]. The work involved in establishing such analyses is extensive. Such analyses will hopefully in the future be of use in dimensioning and optimising health and safety work. There is much need for such optimisation in smaller companies, especially to select the methods which are most effective.

Analyses of the direct impact of management decisions on safety before the event is more problematic. Who would pay for such studies, and how could they be organised? A more fruitful approach is probably to increase management awareness and to solve specific problems rather than using effort to analyse them. The problems are clear enough in any case from reading of the conclusions from accident investigation reports.

REFERENCES

1. N. Rasmussen, *The Reactor Safety Study*, WASH-1400, U.S. Nuclear Regulatory Commission, 1974.
2. P. R. Kleindorfer, J. C. Belke, M. R. Elliott, K. Lee, R. A. Lowe, and H. I. Feldman, Accident Epidemiology and the US Chemical Industry: Accident History and Worst-Case Data from RMP Info, *Risk Analysis*, Vol. 23, No. 5, pp. 865–881, 2003.
3. National commission on the BP Deep Water Horizon Oil spill and Offshore Drilling, The Gulf Oil Disastere and the Future of Offshore Drilling, 2011.
4. The Honorable Lord W. Douglas Cullen, *The Public Inquiry into the Piper Alpha Disaster*, London: H.M. Stationery Office, 1990.
5. U.S. Chemical Safety Board, *Investigation Report: Refinery Fire Incident*, 2001.
6. US CSB, Chevron Richmond Refinery Pipe Rupture and Fire, 2015.
7. OSHA: Phillips 66 Company Houston Chemical Complex Explosion and Fire: A Report to the President, U.S. Dept. of Labor, Washington, DC (1990).
8. Petrobras P-36, Oil Rig Disasters. 14 April 2008. Retrieved 20 May 2010.
9. P-36 Inquiry Commission Final Report, 2001.
10. S. Dekker, Drift into Failure, Franham: Ashgate, 2011.
11. J. R. Thomson, High Integrity Systems and Safety Management in Hazardous Industries, Oxford: Butterworth-Heineman, 2015.
12. R. M.Pitblado, J. C Williams, and D. H.Slater, Quantitative Assessment of Safety Programs, Plant Operations Progress Vol. 9, No. 3, 1990.
13. L. J. Bellamy, Best Practice Risk Tools for Onshore Sies with Dangerous Substances, IBC Conference on Safety Cases, 2000, www.whitequeen.nl.
14. UK HSE, Permit to Work Systems, www.hse.gov.uk/comah/sragtech/techmeaspermit.htm.
15. UK HSE, WL - COSHH essentials for welding, hot work and allied processes, www.hse.gov.uk/pubns/guidance/wlseries.htm.

14 Communication Errors

CASE HISTORY 14.1 Everyday Communication Practices

Charlie and Frederik were unloading rail tank cars filled with inorganic waste from a metal plating factory. The acid waste from pickling (cleaning) steel would be neutralised with slaked lime in a large lined concrete tank. The cyanide waste would be detoxified with sodium hypochlorite (similar to household bleach but more concentrated). Charlie was in the control room. Having received two tanker loads of acid, he observed that the correct amount of fluid had been transferred and that the pump had stopped. He contacted Frederik, who was in the rail yard, by mobile radio: 'Send up another bucket full'. (This sounds more fluent in the original language.)

Frederik disconnected the unloading hose, connected it to the next tank car and started the pump. About 6 cubic metres of cyanide waste had been pumped to the acid tank, before the release of hydrogen cyanide gas was registered and the pumping shut down. (The gas detectors were mounted on and around the cyanide tank, as required by the design, and not on the acid tank, where there was 'no apparent need'.) No one was hurt, but the company was embarrassed, this incident being one in a fairly long series.

After the accident, two separate pipelines were installed, with large and clear signs, telling which kinds of waste would be accepted into which pipe. An investigation was made to see whether tankers could be fitted with different discharge nozzles (this was unsuccessful). A new practice for communication was introduced, with a fixed system of messages and identifications, along the lines of those used in military firing exercise communications or as used by air traffic controllers.

Communication errors figure in many process plant accidents. There is a need for the following:

- Communication between board operators in the control room and field operators, to tell them to take readings, start or stop pumps and to line up (open close) valves for particular plant configurations
- Communication between supervisors in the control room and board operators, for implementation of production decisions, for coordination between maintenance planning and production and in emergencies when supervisors should be standing back to get an overview of the situation while the board operator reports status and effectuates control

- Communication between shift supervisors and between board operator at shift change, to ensure that the status of the plant and the stage in operating and maintenance plans are properly known; also at this time, details of any operating problems, disturbances or failing equipment should be communicated

There are two means of communication for such a shift change. One is direct verbal briefing. For this, a handover period of half an hour is usually planned, where shifts overlap, although a full half hour is generally needed only when there are problems. The second means of communication is the operator log, which should record operations performed during the shift, any change in plant configuration (which tanks are receiving product, for example, or any equipment shut down for maintenance), the plant status and any problems which have arisen:

- Communication from operators in the field to the board operator to tell when tasks have been completed or to report problems
- Communications during emergencies to report status and observations and to make instructions

COMMUNICATION ERROR TYPES

One type of error is simple lack of communication. This is particularly prevalent at shift changes, when operators or supervisors arrive too late or leave too early for a handover briefing. Another is failure to record actions in the operator log, by forgetfulness or simply due to poor reporting skills.

Communications between field and control room may be lacking because of excessive communications traffic, 'dead spots' in the same communications system or noise (both electrical and sonic).

Operators, of necessity, use quite special language. What a piece of equipment is called can vary from plant to plant though. As an example, the vessel used to separate out liquid droplets prior to a gas compressor may be called a scrubber, a knockout drum or a filter! A new operator needs to learn the local names of all the equipment.

Sometimes, new operators may not know the precise equipment referred to by some arcane term. Also, they may be afraid to ask, not wanting to show ignorance in front of new colleagues. Worse, they may think they know or may interpret a term differently. For example, the use of the term *filter* for a knockout drum, as referred to above, can be very confusing, since the term is used normally for a very different piece of equipment.

A good training programme will introduce an operator to all of the equipment items in a plant, their functions, their names and how they work. Such training programmes, though, require careful writing by supervisors and operation engineers, with both long experience and good writing skills. First-class plant-specific training material of this kind is found in only a fraction of process plants around the world. More often, training material consists of some plant-specific material, plus a large quantity of generic information. A good test of training material is to check that it describes all operating modes, disturbed conditions and troubleshooting and that it includes photographs both of the equipment externally, and of vessel internals.

(It can be several years before a new operator actually sees inside a vessel, and internals of items such as cold boxes are rarely seen once they have been built. Good pictures are really needed if the operator is to understand the functioning.)

LANGUAGE DIFFICULTIES

There are special problems in operations when the staffing is multilingual. Some operators will then have difficulty in understanding, and it may be necessary to have a good bilingual operator to check that communication in the standard language is understood. To see the extent of the problem, I asked recently, in a large control room, how many nationalities were represented. The answer—seven if we include you!

It is interesting to see the mode of cooperation in a situation like this. In the actual control room there was a good and very peaceful atmosphere, with notably quiet communication. No one wished to disturb the operators at their workstations. The culture was a mix of nationalities only outside the control room. Inside there was one culture, careful and safe operation.

CASE HISTORY 14.2 A Surprising Problem in Written Communication—Illiteracy

One very experienced (18 years) operator went out to a tank farm and adjusted valves to fill the wrong tank. He operated the wrong valves. In the subsequent accident investigation, it was found that he could not read the work order.

DRAWINGS AS A MEANS OF COMMUNICATION

One item of communication which is very prone to error is that of drawings. If drawings are out of date, there may be additional pipes, valves, etc. which the operators do not know about. Such items may cause unwanted blockages, provide routes for unwanted flows or defeat the isolation during maintenance.

Provision of up-to-date drawings is part of any good management of change (MOC) procedure. However, it is rare to find a company in which small changes are made on drawings just as soon as the physical change is made. For reasons of convenience, several small changes are 'stored up' until a full updating becomes 'worthwhile' or until an audit is expected. This approach is guaranteed to cause an accident at some stage.

Some other drawing office practices are potential accident causes. Engineering office managers do not like unregistered drawings to be kept in a plant, because they are guaranteed to be out of date. Provision of online drawing databases has allowed the elimination of all paper drawings in some plants. However, companies do not always provide good access to the operators and technicians who need the drawings. For example, on one offshore platform which I audited, the Internet bandwidth was so limited that it took half an hour to receive a drawing. As a result, operators and maintenance technicians in such places tend to keep their own 'private' drawing sets, which brings us back to the original problem.

There is no real excuse for the problems of access to drawings in this age. If communications bandwidth is limited, drawing databases can be 'mirrored' automatically on a local server. Out-of-date drawings can be 'red marked' by operators or by supervisors after modifications have been made or as part of the MOC procedure. The red marking can be made on computer-aided design (CAD) systems, without affecting the original drawing, and will be available to all until a full drawing revision can be carried out.

Warning

One warning about as-built drawings: A company, concerned about the state of its drawings, contracted with an engineering company to provide as-built CAD drawings. When these were submitted for a HAZOP analysis, the drawings were found to reflect the originals very faithfully, including drawing errors and erroneous items which had been corrected in the actual plant. The markup of changes over the years was incomplete, and the draughtsmen making the as-built drawings did not make a complete survey of the plant.

SHIFT HANDOVER

Shift handover is a critical aspect of communication. The status of the plant, any abnormal conditions and any problems which have occurred must be transferred from the operations supervisor going off shift to the supervisor coming on, and the information needs to be spread to all operators.

Ideally the handover would be both verbal and written, but staffing schedules often do not allow for this. Shifts may be 8 hours or 12 hours and it might be thought that supervisors could work over for 30 minutes to ensure good handover. This is possible on some plants, but transport arrangements often prevent this. This is especially so in large plants where entry of personal vehicles is prohibited. When you need to go home, you must catch the bus, and if you are late, there are personal problems.

Operators vary widely in their writing skills. Some can write a full page in a log, with an in-depth description. Others have difficulty in making more than a sentence or two. Some companies have introduced therefore a fixed form for handover, in which the information to be handed over is clearly specified.

Lack of communication played a major role in the Texas City explosion of 2005, where no shift handover was made at the shift change prior to the accident. It was also a major factor in the Piper Alpha accident in 1988 [1], where information about the status of the equipment was left as a note on the control station keyboard, rather than on the formal operations log. The note had not been noticed or had gone missing. As a result, a compressor was started with a flange still open.

Various systems exist which should ensure that state-specific information is delivered to the next operator at handover. Examples are shift change handover procedures, entries in the operation logbook, representation of valve status on displays and visualisation of bypass status on safety shutdown instruments and shutdown valves. Prediction of accident frequencies due to lack of knowledge, therefore, involves an investigation of which systems are in use; whether the procedures such as handover briefings really work or have become just an empty routine and assessment of the reliability of the knowledge handover delivery process. It also requires an assessment of how often unusual states arise. As an example, the status of the systems at Piper Alpha had not been reported from one shift to the next; the shift operators never met. The first operator stated that he had left a note on the control panel, a decidedly nonstandard approach to handover briefing.

The probability of accidents arising due to lack of plant state knowledge is therefore the probability of the abnormal situation arising in the first place multiplied by the probability of failure of the handover procedure.

It is the shift supervisor, in the first instance, who needs the transfer of information. Following this, the information must be relayed to operators and maintenance technicians, who will actually effectuate decisions.

Typical abnormal situations which require to be noted are the following:

- Items performing abnormally
- Items of equipment being in a failed state
- Items locked out for maintenance
- Items opened for maintenance
- Items isolated, e.g. by blind flanges or spades
- Persons present in hazardous areas

Failure of verbal handover procedures arising because

- Management does not allow for time overlap between shifts
- Too much is happening; there is simply too much work and information to be transferred within the time available
- Handover procedures have become too relaxed, for example, after a long period in which there are no problems to communicate
- Simple oversight

Reading the operation log can compensate for lack of verbal communication. The extent to which this is possible depends on how well the operator writes, how much time is available at the end of shift and how much the senior operator or supervisor understands the needs of shift change replacement.

Providing formatted handover reports or fixed format operation logs has proved more effective than free-text reporting. Not everyone is capable of literary excellence at the end of a long shift and not everyone can remember all that needs to be said. A fixed format can alleviate the problem, providing spaces for all the kinds of information needed. It is much easier for a person to 'fill in the blanks' than to try to compose a complete and coherent text from scratch.

PERMIT TO WORK AS A SOURCE OF PLANT STATE KNOWLEDGE

In many cases, PTW procedures and lockout–tag-out (LOTO) procedures provide a much better defence against accidents which arise due to lack of plant state knowledge. When designed properly, when PTW offices are properly staffed and where proper LOTO facilities are provided, the reliability of these is very high.

Of the failures of PTW procedures, the following cases give some idea of the distribution of causes of failure.

CASE HISTORY 14.3 Delays in Permit Can Cost Lives

Two contractors entered a vessel and started to remove internals for cleaning and replacement. They did this before the Health, Safety and Environment (HSE) inspector arrived, tested the atmosphere for oxygen content, flammable gas and hydrogen sulphide. The reason was their desire to 'get on with the work' and impatience with the HSE rules. The HSE inspector had been delayed because of the need to solve other problems arising elsewhere on the job. This incident was reported as a near miss.

CASE HISTORY 14.4 A Bad Response to Frustration (see also Case History 11.7)

Two labourers refused to carry out a job of clearing a blockage in the outlet of a hot solution tank until a proper HSE inspection had been made. The foreman insisted they should go ahead, because production was stopped. The HSE inspector, after he arrived, approved the method of emptying the tank by pumping out from the top but did not approve any additional work until the tank emptied and arranged to return later to check the cleaning and the atmosphere in the tank before removing the manhole cover.

A few hours later, the one labourer was found dead, killed by the hot solution, and the other was severely injured. Some of the manhole bolts had been removed; the others were ripped away by the liquid pressure. Either the maintenance labourers had started to open the manhole cover before the tank was drained or possibly they had been doing 'hot bolting', that is, loosening bolts and then retightening them, to make later removal of the cover easier and faster. They were probably unaware of the unzipping phenomenon, which can occur when this is done on a large cover, due to the high stresses placed in single bolts. They were also probably unaware of the high force (about 2 tons) being exerted on the cover by the liquid in the tank. The actual cause of the accident is not known but impatience certainly played a part in setting up the circumstances for the accident and possibly some residual anger that the foreman had insisted on work being done before safety approval was achieved.

In about one-third of the installations audited for a company, work permits were found to have been signed without HSE inspectors visiting the work location. The reason for this was the lack of staffing. Inspectors were fully occupied in filling out PTW forms. While this ensured that the proper checks and procedures were documented, it did not ensure that the workplaces were safe. In other audits LOTO procedures are intended to ensure that electrical equipment, valves, etc., are locked in their safe position while people are exposed during maintenance. One of the most important features is that the people at risk keep the key.

LOTO procedures can fail if there is a single lock, and two persons are involved in carrying out independent tasks out of sight of each other. If one person completes his or her task but forgets or is unaware of the other, he or she will remove the lock and tag, with dangerous results. This problem can be overcome by providing individual locking systems (one lock per person). There are some proprietary lock clamps which allow several locks to be closed, one for every person at risk.

Errors in which blind flanges and closed valves remain closed after maintenance can be prevented by completing 'isolation lists' in which each isolation measure is listed as it is put in place and is crossed off as it is removed. For large units, isolation cannot be carried out without error in the absence of such lists.

SAFEGUARDS AGAINST COMMUNICATION ERRORS

1. Establish one common language for operations and maintenance, and ensure that all have the necessary language competencies. If necessary, provide training. Issue certificates of competency.
2. Establish one nomenclature for the plant, including acronyms. Provide a lexicon of the plant terminology.
3. Make the language form used in radio communications formal. Identify all equipments by tag number and functional name.
4. Establish a procedure of confirmation of any instruction. The receiving person repeats back the order.
5. Do not make unconfirmed assumptions about the understanding of a communication. For new employees especially, confirm that instructions have not only been heard but also understood.
6. Give information about why a particular action is to be carried out. This helps field operators, for example, to understand the urgency, if any difficulties arise, and to be able to communicate back if there are difficulties.
7. Ensure that channels of communication are clear. There should be no dead spots in a plant where radio communications cannot be heard.
8. Provide multiple modes of communication. In plants, telephones can supplement radios when the communication channel use is heavy. Mobile telephones often provide better communication than radios. Provision of explosion-safe mobile phones can help.
9. Ensure that communication can be made to people wearing breathing apparatus and for people inside confined spaces such as pressure vessels.
10. Establish good PTW systems and audit them periodically.
11. Establish proper LOTO procedures and audit them periodically.

REFERENCE

1. The Honorable Lord W. Douglas Cullen, *The Public Inquiry into the Piper Alpha Disaster*, London: H.M. Stationery Office, 1990.

15 Error Syndromes

There are some patterns of multiple errors which are quite complex but which occur sufficiently frequently that they can be regarded as typical error syndromes. Through study of case histories such as those just described, it is possible to identify a number of characteristic error types. These may, in action error terms, seem quite complex, yet nevertheless they recur. They have been summarised here and listed below.

Special Situational and Psychological Error Mechanisms;

- Improvisation
- Shortcut (omission and better sequencing)
- Reversion to stereotype action/standard action/earlier norm
- Omission due to distraction
- Procedure shift due to distraction
- Neglecting a side effect of an action (forgotten, overlooked, simply not known)
- Neglecting a precondition for an action (forgotten, overlooked, simply not known)
- Tunnel vision
- Information overload
- Fixation
- Shift change—omission of procedure
- Change of procedure
- Neglecting a condition
- Reinterpretation of indication as instrument error
- Forget–remember syndrome
- Interchange
- Double charging
- Piling on the gas
- Interchange of objects of an action
- Misinterpretation of ambiguous message or label
- Dismantling active equipment
- Ignoring nuisance alarms and also serious alarms
- Switching off safety systems
- Personalised emergency procedures
- Ignoring hidden warnings
- No allowance made for individual specialties
- Extraneous acts—dropped tools
- Damage while cleaning
- Damage while testing or repairing

Some of these mechanisms will be recognised from the earlier model of operator error. Most are, however, quite specific. They occur repeatedly in accident reports.

IMPROVISATION

When something out of the ordinary occurs, there is often a perceived need to adapt equipment to meet the need. An example is the case of acid leaking from a road tanker. This was pumped over to an available tanker, to stop the leak, but as a result, the tanker contained contaminated acid. Subsequent purification attempts led to an explosion, because the extent of the contamination was unknown.

The incident at Flixborough in 1974 was caused by a similar improvisation—a damaged tank was bypassed, and the bypass was not correctly designed. As a result, it buckled when the operators attempted to use it.

Typical improvisations arise when there is some substance in the wrong place, some standard equipment damaged or missing or a lack of some necessary material. Use of nonstandard piping, nonstandard bolts or 'any pump which is to hand' are special problems. Several accidents have been caused by improvised use of pumps capable of delivering much higher pressure than the piping and vessels were designed to withstand. An error in pressure regulation then leads to disastrous consequences.

CASE HISTORY 15.1 A Questionable Improvisation

An asphalt pipeline was used to conduct asphalt mixed with low-radioactivity waste to a filling station for waste disposal drums. The line plugged with stiffened asphalt. The solution selected by the operator was to pressurise the line with air. This resulted in the asphalt plug being cleared and a column of liquid asphalt being blown out. The operator was severely burned.

SHORTCUTS

It is often tempting to omit or shorten steps in a procedure which do not contribute directly to the success of the procedure but only to its reliability or safety. Waiting for a batch of product to cool is an operation which might be shortened, for example. Subsequent mixing with other reagents can lead to unwanted reactions.

CASE HISTORY 15.2 A Fatal Shortcut

During inspection of a potassium nitrate plant, a water hose was found on the top of a melting furnace, used to prepare the product as a liquid for pumping to a 'prilling tower' where the liquid would flow through a sieve plate and cool into pellets as it fell. The presence of a water hose above liquid potassium nitrate at 350°C presented an obvious steam explosion hazard and should make any safety engineer nervous. The hose was removed. Two months later, a foreman had returned the hose and used it to speed the cooling of the partially cooled nitrate. The liquid was not sufficiently cool, and a steam explosion *did* occur. The foreman was severely burned and died 2 days later.

There can be many reasons for operators to take such shortcuts. Often they are taken from the desire to maximise production or the need to complete a batch before a shift ends or the weekend begins.

There are particular dangers if the operators do not know the reason for a procedure step. Operators will sometimes rail at 'stupid design engineers' who insist on doing things the slow way. Such opinions are reinforced if the operators have experience of actual design errors, so that their opinion of design engineers is low. In some plants lack of respect for the designers is endemically low because 'they don't know anything about operations'. Quite often, the disrespect is justified. Consider, for example, the badly designed valve operation platform in Figure 15.1.

FIGURE 15.1 Example of a valve for which the handle is removable and can give a misleading indication of valve status, open or closed.

Sequencing of operations can often be poorly planned, requiring the operator to walk back and forth between equipments and to climb up and down ladders frequently. Such situations are an invitation for operators to 'optimise' procedures, by regrouping steps. Of course, not all operators will do this. The tendency depends very much on operator experience and awareness of danger. It depends very much, too, on the degree of emphasis management places on safety and the extent to which they are able to communicate their concern for safety culture to the employees, particularly to supervisors.

CASE HISTORY 15.3 Improvisation Forced by Inadequate Design

The author was asked to develop a procedure for isolation, draining down and cleaning of a solvent distillation unit. There were some drain valves, but large parts of the piping could not be drained because of lack of valves at many of the low points. Many of the drain valves were also out of reach and required climbing on structural steel or installation of scaffolding in order to reach them. The draining down was made by first flooding the system, to flush out as much solvent as possible, draining as much as possible through drain valves and then opening flanges to get the rest. This was still dangerous, because the flushing could not be perfect, and this necessitated extensive fire precautions. After successful cleaning, the number of drain valves on the unit was doubled.

Quite often there are steps in a procedure which are not critical for obtaining the main objective but are important for safety. Operators wishing to speed up tasks may leave out these safety-critical steps. Even more likely, operators who do not know the reason for the steps may omit them.

Case History 15.4 shows the importance of explaining not only what to do but also why in written operating procedures.

CASE HISTORY 15.4 An Expensive Shortcut

After one production a kettle-type batch reactor was to be allowed to cool, before adding raw material and solvents for the next batch. An operator tried to accelerate the process, shortening the cooling time. The enamel lining of the reactor cracked, requiring an extensive repair.

Respect for written operating procedures is often weakened by the presence of errors. These arise sometimes because the engineers who write them do not know everything needed about the plant; because they are copied from other procedures for slightly different designs or because the plant design does not work or has been changed.

CASE HISTORY 15.5 Improvisation Forced by Another Design Error

An LPG storage farm was designed with too little height difference between the vessels and the transfer pumps. As a result, the pumps cavitated or became gas bound each time they were started. Operation staff learned to operate the pumps by venting gas from them during start-up, an operation producing a release which would normally be regarded as a serious incident.

The case indicates the way that respect for procedures, for engineers and for safety can be reduced by a simple design error.

In some cases, procedure and safety regulations are written in a way which is impossible to perform.

ABANDONING THE TASK

There are many cases of incidents in which an operator starts a task and then departs, to do something else. These are usually tasks which involve filling, emptying, heating, cooling or reacting, i.e. tasks which take some time. The reason for leaving the task may be to do other work, to smoke, to take refreshment or for 'comfort calls'. Examples are given in Chapters 3 and 6. The error arises if the operator fails to return to the task sufficiently quickly.

REVERSION TO STEREOTYPE

If a control is unusual, it may well be possible to train operators nevertheless to use it reliably. However, under stressed circumstances, the operator may well revert to stereotype. Examples are operators trying desperately to loosen a left-handed threaded screw, by turning to the right.

Nonstereotype designs involve left-handed screws and instruments which show increases by moving left, down or counterclockwise. Engineers' desire for symmetrical control rooms has in the past been a cause of having two standards for arrangements of instruments.

Designers should also be aware that stereotypes change from culture to culture. For example, many languages are written from right to left, and there is then a clear tendency to misinterpret horizontal indicators.

While it is fairly easy to detect nonstereotype equipment, the analyst should also be aware that the operator might have norms of action which have been learned earlier. Particularly, the operator may have learned other operation sequences, or limits for action, on other equipment. These can resurface in times of stress, such as in emergencies.

OMISSION DUE TO DISTRACTION

Omission of acts due to distraction, or interference of other tasks, is always a potential problem. There is no real way of preventing distraction – virtually all operators have to be available on the telephone, have personal needs and have to split their time between tasks.

Special distractions arise, especially during emergencies, if many other people enter control rooms. High-level engineering and administrative staff can distract operators with questions or just by their presence. Error rates rise during such circumstances.

PROCEDURE SHIFT

It is possible for operators to lose the context of their work, particularly if the work is not very demanding and they are thinking of something else. If they have left the operating room, or left responsibility to someone else for some time, they may not appreciate the situation on return. As a result, they may take up some completely different activity than that in which they were previously engaged.

A special problem here is handover of responsibility between two people. A may not realise that a change of situation has taken place. Unless B is directly aware of the problem, he or she will not realise that A is unaware of the shift. As a result, two operators work on two different sets of assumptions.

NEGLECTING SIDE EFFECTS

Operators may frequently oversee a side effect of some of their actions.

An example, which has occurred many times, is that an operator takes a sample or opens a drain without using the proper procedure, for example, removing pressure from the plant or closing a separate shutoff valve. By careful manipulation, 'cracking open' of a valve, he or she may be able to take a sample. But if there are crystals, tramp metal or crud in the sample flow, the valve may then be blocked open. He or she may be unable to reach the correct shutoff valve, and the drain will continue to leak.

CASE HISTORY 15.6 Side Effects Neglected—Evaporation Cooling Causes Freezing

A sequence something like this occurred in the Feyzin disaster, where a drain valve on an LPG storage sphere froze open due to propane evaporation. The result was a major release, a fire and a subsequent boiling liquid expanding vapour explosion which killed 17 firemen.

An even more direct example of overlooking side effects occurs when a maintenance team receives instructions to overhaul or test a pump or motor, etc. Before starting, they should check that the pump is not required for standby purposes. In several cases, pumps have been taken out of operation, and plants have, as a result, shut down, because the standby pump failed.

CASE HISTORY 15.7 Side Effects Neglected—Valve Blocked Open Due to Foreign Object

In another incident, a very slow leak of liquid sulphur through a blocked open sampling valve caused a process room to be filled with sulphur to a depth of a few feet. The sampling valve was cracked open while the sulphur condenser was still pressurised. A tray screw lodged in the valve trim, so that the valve could not be closed. The operator filled several buckets with sulphur while trying to close the valve but eventually had to give up, and the sulphur flowed to the floor. The plant was shut down, but the remaining inventory in the condenser continued to flow. Once stiffened, the sulphur had to be removed with the aid of pneumatic drills.

In cases where side effects such as starting a motor or applying electricity could have wide ranging side effects, it is usual to have a careful system of permits and tagging of equipment before operation is allowed. Such permit systems are not always perfect, however, and do not always take into account the full range of side effects which can occur. Procedures such as action error analysis can be used to validate permit procedures (see Chapter 18).

OVERLOOKING A PRECONDITION OR LATENT HAZARD

Many actions in an operating procedure are conditional on a whole series of preconditions being fulfilled. The operation should not be carried out until the precondition is ensured. Checking of preconditions is often an 'unnecessary' part of a procedure, in that in most instances, the procedure can be safely carried out without checking. In this case, only in seldom circumstances will operation without checks be dangerous; such situations are especially error prone, because when a need for checking does arise, it is unlikely to be carried out.

CASE HISTORY 15.8 Pumping Out a Tank without Checking Venting

A classic example of this, which has been repeated in most process plants around the world at some time, is to begin pumping from a tank before ensuring that there is an adequate flow of air into the tank. Ventilation lines may be blocked because of valve closure or because of ice, waste, etc. One common cause of inadequate ventilation is that painters have tied a plastic bag over vents (in order to avoid being affected by fumes). Another is that birds have nested in the vent pipe (I found an actual case in one plant). For most storage tanks the result is that tank walls are 'sucked in' and the tank collapses.

It is not even necessary in such instances for a pump to be started. If the outflow line from the tank has a big enough fall, the vacuum generated by the falling liquid alone can be enough to damage a tank.

CASE HISTORY 15.9 Overlooking a Sneak Flow Path

Another example arose when an operator wished to replace a flow-regulating valve in a liquid chlorine system. He chose to make the replacement in a 'rapid' way and closed block valves at either side of the regulating valve. He was aware that these might leak and therefore closed the block valves very tightly. What he was unaware of was that a burst disk bypassed one of the block valves and served to prevent the pipe section from overpressuring with liquefied gas. The burst disk had in fact failed. On unbolting the regulating valve, this latent condition became apparent in the form of escaping liquid gas. Even then, the leak appeared only to be a bleeding off the contents of the regulating valve until the valve was totally unbolted. Then the gasket which had stuck to the flanges came loose, releasing a flood of liquid chlorine, in all 80 kg.

TOO NARROW BASIS FOR A DECISION: TUNNEL VISION

Operators faced with a mass of instrumentation often select a few indicators as primary indications of the state of the plant and its subsystems. With sufficient experience, they come to base their decisions on these 'reliable indicators' alone. This means that they may well neglect other indicators which could give them useful information during disturbances.

More common than this is the tendency to look at single indications at a time and not to relate indications to each other. During the Three Mile Island incident, there was sufficient information in the form of temperature and pressure indications to reveal that water in the reactor core was boiling. However, this information was not put together and conclusions were not drawn until very late in the course of the incident.

FIXATION

If a disturbance has occurred, then one of the first natural reactions is to investigate possible causes. If a hypothesis has been proposed and accepted, then there is often a tendency to cling to this hypothesis, even after there is positive evidence that the hypothesis is wrong.

The incident at the Three Mile Island reactor provides an example. Once the operators had hypothesised that the reactor was 'overfull' with water, they stopped emergency core-cooling pumps. They retained their worry about overfilling the reactor, even after the water level in the reactor was excessively low.

If an assumption is made, on purely hypothetical grounds, by one person in a group and the assumption fits some observed facts, others in the group may regard the assumption as proved. Such an effect can be self-reinforcing and can lead to diagnoses with no foundation in fact at all.

SHIFT CHANGE ERRORS

Shift change is a situation during which errors often arise. At shift change, one operator must inform another about plant state and about the stage reached in a production schedule. Anything forgotten during this orientation or expressed ambiguously or misunderstood, will provide a cause for later problems.

THE FORGET–REMEMBER SYNDROME

If operators forget a particular step in a procedure, this may not have any immediate effect. However, if they later remember their omission, it is only natural for them to hasten to correct their mistake. For some plants, the correction can be disastrous.

A particular situation of this kind is that of applying heat to a batch reactor, prior to addition of a reagent. If heating is applied, the slow addition of reagent will be accompanied by production reactions which prevent the reagent from accumulating. If heating is forgotten, the concentration of reagent may, on the other hand, gradually build up. The late addition of heat can then lead to a runaway reaction.

CASE HISTORY 15.10 Forgotten by the Operator, Remembered by the Foreman

In a soya oil factory, crushed soya beans are fed to an extractor, a large vertical cylindrical vessel, with rotating trays. The soya meal falls from one tray to the next while hexane is sprayed over the meal and steam is passed upward. The result is that soya oil is washed out of the meal. The hexane runs to the base of the extractor and is then pumped away, to be distilled, so that the hexane is completely removed, leaving purified soya oil.

A new operator started up the plant, with little supervision. After about 2 hours, his supervisor returned and observed that the steam had been started, but not the hexane. He immediately adjusted the hexane flow to its normal level.

By this time, the extractor was very hot, with a temperature much higher than the boiling point of hexane. The hexane flashed, pressure rose and a bell-type safety valve blew open.

Hexane continued to escape for about 1 hour, until ignition occurred, with a large unconfined vapour cloud explosion.

CASE HISTORY 15.11 To Start or Not to Start?

In a bromination plant a slow addition of bromine into a reactor was started, but the operator forgot to start the reactor agitator. After some hours, the omission was noticed. The operators were aware of the danger, but after discussion, decided to start the agitator anyway. A runaway reaction occurred. A burst disk on the reactor opened, and bromine vapour escaped. The operators were trapped in the control room for several hours.

Forget–remember situations can be readily identified from the key word *forget* or *omission* in an action error analysis. If forgetting leads to an accumulation of energy by a hazardous substance, then there is a possibility of a serious accident when the omission is finally remembered. From examination of incident reports it appears that the delayed remember error occurs in about 50% of the cases where it could be hazardous.

REPAIRING THE WRONG ITEM, DISMANTLING THE WRONG EQUIPMENT

Identification of items to be repaired is a frequent problem when there are several alternatives. Almost every year, accidents occur because maintenance teams dismantle working equipment. The cause may be misinterpretation of instructions, wrong instructions, lack of coordination or ambiguity in instructions. More often it arises because two similar production units are confused.

If flanges are partially unbolted, then equipment under pressure can, and sometimes does, break the remaining bolts and cause pipes and vessels to fly apart. Opening doors behind which pressure has accumulated can cause similar violent mechanical accidents.

CASE HISTORY 15.12 Dismantling the Wrong Pressure Safety Valve

Pressure safety valves (PSVs) are often fitted in pairs, with interlocked block valves, so that one valve is always piped up and ready to open. This allows one valve to be dismounted, tested and reconditioned if necessary, without needing to shut down all production.

In the actual case, the fitter blocked off one valve on an ammonia vessel, then dismounted the other. Ammonia escaped, and the fitter was killed.

This case history is one which is typical of many. There are several published cases of such errors for ammonia vessels alone. The frequency increases if the block valves are not located close to the safety valves; if three-way valves are used, with no clear position indication and if the piping is complex with poor labelling. Case History 6.8 provides another extreme example of this.

Identification of potential for mistaking equipments is relatively easy – if there are two possibilities, it is almost certain that at some stage the wrong one will be taken. Probability of error will be greater if any of the following occurs:

- There is no clear labelling.
- There is no indicator to show which equipment is operational.
- Piping is complicated.
- Equipment arrangement is not clear and logical.
- The position of a valve is unidentifiable or illogical.

Examination of case histories shows that there were no reported cases of such errors in the database unless there was at least one of the above features present.

Given error-prone features in the design, the probability of error depends on experience, either of operating the actual equipment or of the need for caution. Without experience, the probability of error prior to the accident, as determined by checking how often the tasks were performed correctly, varied between 0.03 and 1.0.

THE EAGER BEAVER EFFECT

Some workers, mostly new employees, are eager to do well. This can cause them to carry out well-intentioned but dangerous actions.

CASE HISTORY 15.13 A Novice Employee at Risk

A person had been employed just 2 weeks and was helping to install a new conveyor. The conveyor belt was being pulled into place by a bulldozer (it was heavy). The cable used for the pulling was attached to the belt by a heavy cleat.

Unknown to the supervisor, the man climbed to the location of the belt. He had just enough time to say, 'Everything is OK', on his radio when the heavy cleat was torn from the belt and flew through the air. It hit the man in the skull, killing him immediately.

He was eager to help, but he knew too little to stay safe. The supervisor was without blame; he had instructed everyone to stay clear.

CASE HISTORY 15.14 Heroic and Selfless, but Wrong

In an iron foundry, crucibles are used for transporting liquid iron from the furnace to the moulding floor, where the iron is poured into the moulds. As time goes on the fireproof brick lining of a crucible will erode away and must be replaced. Sometimes this does not occur early enough. The iron then eats its way through the lining and melts/dissolves the steel casing of the crucible. Liquid iron then escapes. If the iron pours onto wet sand, an explosion can occur.

In the actual incident a moulder noticed that a crucible casing was glowing red hot. The crucible stood on a railcar. The moulder decided to push the crucible back to the furnace area, where spills of iron would not be dangerous.

The railcar got stuck on a section of track where it was connected with fishplates (strips of steel used to link tracks) due to unevenness in the rail. The iron escaped from the crucible and ran into a cable trench. The iron flowed all the way into the control room (at a lower level) and ended in the computer system. No one was hurt, but the furnace controls were unusable for months.

Note: If you succeed in this kind of action, you are a hero. If you fail, you are just someone prone to dangerous foolhardiness.

PERFECT TRAPS

A perfect trap is one in which the operators have every right to feel that the plant is safe and that they can proceed with normal operation safely. This may be because all the safety indicators are showing safe, because they have been told that the plant is safe, and because they have seen the plant being checked for safety. Nevertheless, when they begin the task, a latent hazard is activated and may kill.

An example occurred when a maintenance fitter was required to replace a pump. The suction and discharge valves were 'closed' and the drain valve showed no flow into the open drain. In fact, the suction valve handle had been taken off and refitted with a 90° change in position. This meant that the valve indicated open when closed, and vice versa. Also, the drain line was plugged with tarry residue, so that the fitters could not see that the pump was still pressurised. When the flange was opened, hot naphtha was released. The vapour spread into the plant and caused an explosion and fire. Six employees were injured [1].

Falling into a perfect trap is not a human error. Creation of a perfect trap can be a human error, such as replacing the valve wrench wrongly in the example in Figure 15.1 above. Even then, such errors may be the result of lack of knowledge and instruction, so the error devolves back to management.

CASE HISTORY 15.15 The Instrument Works Perfectly until You Really Need It

A temperature alarm was to be fitted to a chlorine evaporator, to indicate a low temperature which might cause the heating fluid, glycol brine, to freeze. In practice, the alarm was fitted to the brine discharge pipe from the evaporator, so that it worked well while the brine was being circulated but failed completely when most needed, that is, when the brine pump failed. During commissioning, a power failure occurred, the pump stopped, the brine froze due to chlorine evaporation and piping cracked. As a result, chlorine escaped.

CASE HISTORY 15.16 Another Perfect Trap (Figure 4.2; Ref. [2])

A distillation column was fitted with a level switch, to indicate high level in the column bottom section, in which the less volatile component of the feed fluid collects. The alarm indicates that insufficient product is being pumped from the column (or too much product is being fed to it). If the alarm signal is given, it is important that the feed is reduced or the pumping out increased, to prevent the lower trays of the column from flooding. Flooding can lead to tray damage.

In the actual incident, the alarm was ignored during plant start-up, because it had always occurred until the column could be brought into balance between the rate of feed and the rate of heating. In this case, however, the operator had forgotten to start the column bottom's transfer pump, so that the column continued to fill until it overflowed. The PSVs opened and the product (naphtha) flowed to a vent system, eventually spraying from the vent and causing an explosion.

CASE HISTORY 15.17 Trapped Pressure

A pipe had gradually built up a layer of silica until at last flow stopped. A maintenance team first opened all drain valves on the pipe, then unbolted the flanges carefully and removed the pipe spool. During the removal, the blockage in the pipe began to leak, spraying dilute acid. The pipe fitter was wearing safety glasses, but these were not adequate to protect his eyes as acid ran behind them. Fortunately, the safety shower and eyewash were close by.

REINTERPRETATION OF INSTRUMENT FAILURE AS NORMAL

Alarm systems are intended as an indication to the operator that something is wrong. High and low alarms are provided at disturbance levels which allow the operator time to adjust the plant and avoid the need for shutdown. 'High high' and 'low low' alarms are provided to indicate that the operator *must* shut down, although more often in modern plants, they are coupled to automatic shutdown systems.

When an abnormal indication is shown on a display, the cause may be either a plant disturbance or an instrument failure. The operators must decide between these alternatives. They will be especially prone to doing this if their experience includes more instrument failures than plant disturbances. Such is the case for much instrumentation intended to monitor for emergencies. The problem can be so extreme that when an emergency arises, the operator recalibrates the instrument.

Spurious activation occurs if there is a failure in the instruments or the electronics for the alarm or if there are disturbances in the plant which give an alarm level without being serious. Typical examples are the following:

- Vibrations occurring during plant start-up which quickly die away
- Pressure pulses arising during pump start
- Level sensor disturbance due to foaming or to splashing waves when liquid transfer is started

Spurious alarms reduce belief in alarm integrity. From interviews done after control room observations, just a single spurious alarm will reduce belief in a sensor.

Instrumentation system design has changed a good deal in the years from 2000 especially, when high-quality safety integrity level (SIL)–rated equipments became available at reasonable prices and the HART system for instrument self-testing became standard. This has made instruments much more trustworthy. A shift to indicating instruments rather than alarm switches has also meant that the performance of instruments can be followed in the control room by the operator. A flat line on a trend curve for a valve which should be varying, for example, is a clear indication that an instrument is stuck.

Even with these improvements though, there are sometimes features in instrument design which function as perfect traps for operators.

Instruments with a limited measurement range are frequently necessary in process plant design but can easily mislead. If the instrument indicates that, for example, level or temperature is high, it is natural to interpret the temperature as being the 'high' instrument reading, when in fact the value may be much higher, perhaps catastrophically high.

An example of this which is so frequent that it can be regarded as a standard case is that of hydrogen sulphide gas detectors. These usually have an upper reading limit of 50 parts per million (ppm) or 200 ppm, which is a very dangerous but survivable concentration provided that you are aware of the cloud and get out of it quickly. The instruments give the same reading, though, for all concentrations up to 1,000,000 ppm (except that some instruments saturate and fail to give a reading at all at these levels).

Identifying situations in which instrumentation may be misleading or mistrusted is quite straightforward during HAZOP analyses or especially as part of SIL reviews. Also in human factor studies, review of individual instruments performance is possible. A checklist is a useful aid in these:

Checklist Items for Instrument Integrity:

- Does the instrument register the actual hazard, or is the hazard a secondary effect?
- Does the instrument register the hazardous parameter, or is the reading a secondary effect of the actual hazard?
- Will the instrument work under all operating conditions?
- Are there any situations in which the alarm signal will occur normally? If so the alarm is likely to be ignored.
- Can the instrument be bypassed? If so, is the bypass condition obvious and will the bypass conditions have a limited duration?
- Are instrument failures automatically detected? If there are some which are not, is the failure obvious to the operator? (e.g. with a flatliner on a trend indicator)
- Is it possible for the operator to see the alarm instrument reading, so that the alarm can be reset?

Dismantling blocked pipes is a frequent source of perfect traps; blockages in the pipe or closed valves can trap pressure or hazardous substances.

INFORMATION OVERLOAD

Information display systems work well while the plant is in normal operating state, or with minor disturbances. In large process plants, serious disturbances often generate display information at a rate which is far beyond any operator's ability to comprehend or to respond to. Individual operators can respond to two or three alarms requiring their immediate reaction. They can interpret patterns of up to five or seven alarms or abnormal indications [3]. Beyond this, they must use various serial search and priority strategies. Displays must be very carefully designed, and the operators must be trained in response to each form of emergency if they are to respond quickly in the face of more information than this.

IGNORING 'NUISANCE ALARMS'

Many alarms arising in a plant are nuisance alarms, which are the direct and expected result of start-up or other routine operations or which are responses to small events such as slight leakage of toxic gas when a sample is taken. The operators become used to such alarms and ignore them. If such alarms come to be ignored when expected, they can also come to be ignored under unexpected circumstances.

This effect is so extreme for gas-detection systems that plants usually have a rule: Shut down on *confirmed* gas detection. Confirmation may be made by use of closed-circuit television, by having several detectors at the same location, by checking several alarms displayed on a map-type mimic display or by sending a field operator to investigate the leak.

SWITCHING OFF SAFETY SYSTEMS

Safety systems such as cooling systems, pump-out systems and dump-and-relief systems may be switched off by operators fearing side effects of activating the safety system or feeling that the safety action is incorrect. This will especially be a problem in cases where there are conflicting safety goals. It may also be a problem if operators have experience of earlier design errors or problems in safety systems. In Case History 13.5, the operators did not want to activate the relief system because it had earlier resulted in burning rain and grass fires.

PERSONALISED EMERGENCY PROCEDURES

Operators with long experience in a plant may get to know its behaviour better than the plant's designers. They may distrust safety procedures and feel they can do better. This is a good thing provided the revised procedures are discussed in safety committee meetings and formalised. Sometimes, though, such communication is difficult and can lead to conflicting safety responses.

HIDDEN WARNINGS

If safety indicators or alarms are covered by logbooks, coffee cups, hanging labels, safety tags or even just another operator sitting on the workstation table, they may well be ignored.

INDIVIDUAL SPECIALTIES

The problems here concern particularly operators who are left-handed, colour blind or partially deaf. If allowance is not made for these, errors may readily result through misunderstanding.

PILING ON THE GAS

In pumping liquids, the first sign of a major pipe leak or a rupture is that the pressure falls at the pump discharge. An operator who has not been trained for emergencies (and in some cases, ones who have) will respond by curing the symptom, not mitigating the consequences or shutting off the cause. This is not so surprising; pump and valve failures are relatively frequent. Most pipeline operators do not experience a pipe rupture in their entire working life. The response, though, is unfortunate. Increasing the pumping rate into a ruptured pipe makes the accident several times worse.

Other examples of driving process plants hard are seen regularly, with full knowledge, not in error. Figure 15.2 shows a very cold flow from a chlorine evaporator which was driven beyond its design limit. This did not affect the process, but care is needed in this kind of operation to ensure that the temperature does not fall below the level at which transition to brittleness takes place in the steel.

USING SAFETY DEVICES AS CONTROLLERS

If filling of a tank takes a long time, it is tempting, and sometimes necessary, to undertake other tasks. If there is a filling pump trip or a level trip which closes an inlet line valve, any overflow accident will be prevented, that is, until the trip fails.

Using instrumented trips as control devices is a relatively common practice. It usually starts when an operator observes that he or she has forgotten to monitor level or temperature and that the system has worked properly. It is made more likely if workload is high and if the trip simply stops a pump or closes a valve, etc., without shutting down a significant part of the plant or triggering an area alarm which leads

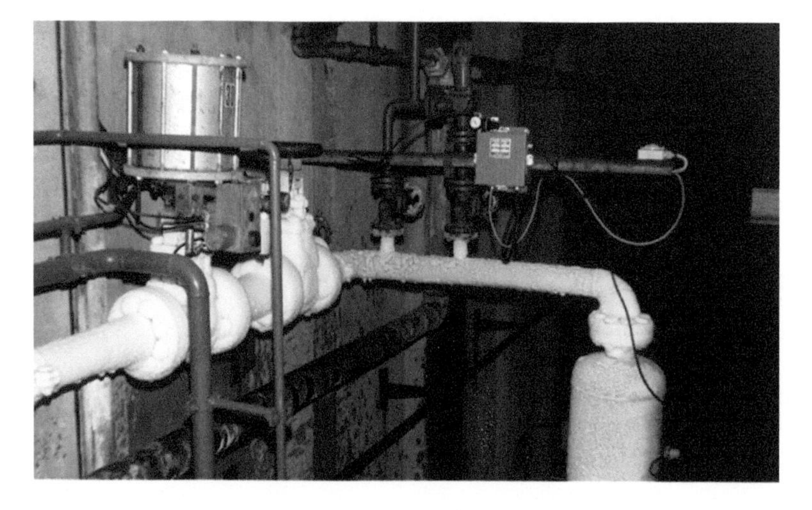

FIGURE 15.2 Low temperature caused by running a chlorine evaporator beyond its design limit. Frost is forming on the pipe. In such cases care is needed to ensure that the temperature does not fall so low that the steel becomes brittle.

to an evacuation. It is also relatively common for the use of trips as a control to develop from an occasional practice to a routine one.

The underlying problem here is that the trips are intended as a safety device and have a probability of failure on demand of typically about 0.01, or 1 chance of failure in 100. The operator who is monitoring the action is generally much more reliable than this, and the combination of operator vigilance and a backup safety device is extremely reliable. That this is true can be observed by the fact that most tank yards are relatively clean, with few indications of overflow, such as streaks down the sides of the tanks.

INSTRUMENT BYPASS

In many plant operations, it is necessary to disable safety devices for a period. Typical examples are the following:

- In start-up of rotating machinery such as compressors, at certain speeds there is resonance between the rotation speed and the whirl vibration frequency of the shaft, or other equipment-resonant frequencies. It is necessary to pass through these speed ranges quickly to avoid machine damage. During this period, vibration trips need to be switched off, because otherwise the machine would shut down automatically and could never be set in operation.
- Sometimes a safety instrument will be subject to intermittent failure. It should be replaced, but if this is impossible, such as due to unavailability of a spare, the instrument will need to be bypassed.
- Sometimes it is necessary to bypass an instrument in order to allow testing of others.
- Sometimes it is necessary to bypass an interlock in order to be able to shut down a plant, to depressure and drain it, after an incident. This is especially the case with sequential controls which allow valves to open only at the correct point in time in a production sequence. Interruption of the sequence then results in the valves being locked closed.

In some cases, bypassing is all too simple. It can take the form, for example, of closing the valve for a pressure switch.

In modern instrumentation practice, most of these problems can be avoided by the instrumentation engineer taking them into account in the design. Nearly all plants still need bypass of safety instrumentation, however, to allow for start-up transients or for testing purposes.

If bypassing is done, it should be done under strict control. Best practice is to require approval by the plant manager, which is given only after risk assessment and only for a short period.

Unless rules for bypassing are very strict, operators quickly lose respect for the safety system, and bypasses are introduced for convenience rather than necessity. An even worse situation occurs when the bypass is forgotten.

Modern instrumentation allows for checking on bypasses. Some controllers have bypass facilities built into the computer-based control system which require a password, automatically register the bypass and send a message to the plant manager if the time limit for restoration is exceeded.

The use of transmitter instruments rather than switch-type instruments for alarms and trips also reduces the tendency for leaving bypasses in place, because it becomes possible to see at all times whether the instrument is functional.

CLEANUP AND CLOSEOUT

When a task is completed, there will generally be a whole range of closeout and cleanup tasks, such as restoring instrument valves to the operational position, collecting tools and replaced parts and washing down drained fluids.

It is much easier to forget these closeout tasks than to forget actions in the main job, because these tasks are not related to the prime goal motivating the work.

USING ERROR SYNDROME PATTERNS IN ERROR ANALYSIS

Some attempts were made in the 1980s and early 1990s to develop error-identification methodologies based on error patterns [4]. These were not very successful. The best approach found so far is to learn these error patterns and to remember them or use a checklist when carrying out other error-analysis methods such as action error analysis (see Chapter 18).

REFERENCES

1. U.S. Chemical Safety Board, *Case Study: Oil Refinery Fire and Explosion, Giant Industries*, 2004.
2. U.S. Chemical Safety Board, *Investigation Report: Refinery Fire and Explosion and Fire, BP, Texas City, Texas, March 23, 2005*, 20 Mar. 2007.
3. Abnormal Situation Management Consortium, *Effective Alarm Management Practices*, Abnormal Situation Management Consortium, http://www.asmconsortium.com, 2008.
4. J. R. Taylor, *A Background to Risk Analysis*, Risø National Laboratory, 1979.

16 Design Error

In Chapter 1, statistics on the frequencies of different causes of accidents in process plants are shown. Design error was found in studies to be involved in 35% of nuclear power plant accidents and near misses (in the 1960s and early 1970s), in 67% of chemical plant accidents (in the 1980s) and recently in 25% of the largest accidents in Marsh's *The 100 Largest Losses*. Actually, these statistics cover only the direct design errors or weaknesses, since over 70% of accidents involve less than adequate safety equipment [1–5]. For nuclear plants the percentage was 46% for safety-significant incidents. Kidam [6] in studying the Japanese Failure Knowledge Database [7] found that approximately 79% (224 out of 284) of accident cases involved at least one design error.

THE DESIGN PROCESS

To understand design error, it is necessary to understand the design process. At the outset of design of a major process plant, there will be a specification or conceptual design stage. The starting point will be a certain need, for example, the need to develop an oil field or to provide new capacity for a plastic production plant. Today the need for refinery or petrochemical plant construction is rarely pure greenfield; it is usually alongside an existing plant.

The conceptual design of the process will be by selection from a range of available processes, all well known in the industry, and documented for those needing the information in the process catalogue published by the journal *Hydrocarbon Processing*. Completely new designs in the oil, gas and petrochemicals industries are rare and are generally made by process development organisations, who then act as process licensors. The reason for this is straightforward—developing a process requires laboratory-scale development, pilot plant construction and testing. The development can take 10 years or more. By contrast, purchasing a set of standard process designs can take as little as half a year. Much more creative and varied process design is made for a fine chemicals plant and a speciality plant such as that for catalyst production.

When the process concept is clear, process calculations will be made in order to determine process conditions and vessel sizing. This is followed by development of a layout, which is one of the most creative aspects of the design. It is necessary to fit all the parts of the plant together with adequate spacing for access and for maintenance and to provide as much separation as possible between the most hazardous parts of the plant and people and the most valuable plant sections. Work is generally needed to fit the design into available space and to ensure a good 'pattern of flow' through the plant so that pipe lengths are minimised. In all, there are about 20 criteria for layout optimisation.

In modern practice risk assessments will then be carried out. Then piping and instrumentation design will be carried out in just sufficient detail to allow a cost estimate of the plant to be made.

In some cases conceptual designs like this may be made by potential bidders for the engineering contract, in which case the plant will be the engineering companies' standard designs, which may be completed in considerable detail.

During the conceptual stage there is an opportunity to maximise safety by adopting inherently safe design principles and choosing layout to maximise safety. After the conceptual design is completed, there is less chance for this, but overall safety optimisation can still be done in the first part of front-end engineering and development (FEED).

Conceptual design may be made by the purchasing company's engineering or project department or by a specialist consultant. The work in the FEED stage will generally be given to an engineering design contractor, on the basis of bidding.

In FEED, the process design is modelled, optimised and fine-tuned. Various selections for alternative designs still left open during conceptual design may be made.

In modern practice, during FEED, design philosophy documents will be written (or copied from various company libraries). These include safety design philosophies. The documents are important for starting the design on the right track. The general content of the documents may have been set down during conceptual design.

The process flow diagrams will be refined and retested, and on the basis of these, piping and instrumentation diagrams (P&IDs) will be developed. Designs of utilities for the provision of plant water, air nitrogen steam and power will be revised, as better information concerning the needs of the process are refined. Equipment-dimensioning calculations will be carried out again, and instrumentation and power and utility supply will be designed.

A good many checks are usually made on the design at this stage including design review, HAZOP study [8], HAZID [9], quantitative risk assessment (QRA), blast analyses, buildability analyses and possibly human factor analyses. In the newest design processes, human factor review, inherent safety review and systematic lessons learned analyses may be carried out. These are important documents. Safety and environmental acceptability is generally a criterion for plan approval by shareholders, banks that provide the capital and authorities.

Finally, the plant cost is estimated. This is an important step, because the budget for the construction must be approved, and any error will result in an embarrassing problem when the bids for the plant construction arrive.

It is important during the FEED stage to get the design well specified, to avoid any outstanding design issues, especially conceptual issues, and to make the design as well tied down as possible. The reason for this is that design changes made at the FEED stage can be made at the cost of often just a few man-hours' work (though often requiring much more expensive committee meetings). Once the design is issued for engineering, procurement and construction (EPC), bidding any changes will be expensive. EPC companies sometimes even provide bids which on paper will result in a loss to them, expecting to make up for the loss in change orders. A FEED design which is incomplete or which contains errors is more or less a guarantee of budgetary tears later.

The EPC design will repeat much of the P&ID design and will add designs for equipment vendor packages. Part of the EPC process is to obtain good bids from vendors to meet the requirements. The EPC contractor will generally have contacted vendor companies for important items such as compressors or packaged items such as water filters and nitrogen plant as part of the preparation for bidding. They may even have selected a package vendor as a cobidder. Sometimes the purchasing company will specify an equipment vendor. This especially covers instrumentation and control, where the purchasing company may need to ensure conformity of new pieces of equipment with existing installations at the site, which may even be operated from the same control room. The design at the E stage of EPC will also produce detailed layout drawings, piping drawings and equipment specification sheets.

Long before the detailed design is complete, work on procurement will have begun, especially procurement of long–lead time items, such as large vessels, compressor, large pumps and blowers, and equipment in the power-generation plant. Also, construction will begin with civil works and foundations long before detail design is complete.

From FEED through to detailed design, construction and commissioning may take as little as 2 years for a small plant; 4 years is more normal.

In all, the process is choreography of large numbers of people in different organisations (often 50 or more) working according to rules developed over years, generally in a very disciplined way, with the discipline enforced by terms of contract, schedule promises and deadlines, agreed standards and the promise of payment and profit.

The actuality sometimes follows the ideal described above. Problems arise when there are misunderstandings; there are ambiguities in requirements; there are persons involved who are less competent than the job requires; competent persons cannot be hired; the cost of materials increases due to tight markets, leaving EPC bidders in a cost bind and egos enter the design equation. The frequent results of upsets in the design process are very tight designs, to minimise cost and schedule delays. These can have important effects on plant safety.

CASE HISTORY 16.1 Penny-Pinching Design

In an ammonia plant, in a plant safety audit, an area was found which had been cordoned off due to a leak from an ammonia transfer pump. Usually, two such pumps are provided for each ammonia storage tank in order to allow for maintenance and repair. In this case one pump had been provided for two tanks. The company could not afford to shut down the plant in order to repair the pump (at a cost of several million dollars for lost production), and were still waiting for approval and reengineering in order to upgrade the design. The original plant had been built to a very tight budget, because the contractor had been working effectively under duress, under a contract which was not wanted.

It is situations like this which are frustrating for operators and can rapidly lead to improvisation. To add an additional pump with no engineering work could be carried out by a maintenance department in less than a week, and the tie-in could be made by 'hot tapping', that is, welding in piping connections with the plant still under operation. Hot tapping accounts for about 0.5% of accidents in the U.S. Risk Management Plan (RMP) database of 5-year accident histories.

DESIGN THINKING

The way in which designers think is important for understanding errors. A HAZOP study is a very good way of observing this, since a complete team of design specialists are generally assembled and discuss designs in detail. The observations below include feedback from some very large HAZOP projects facilitated by the author and from process health, safety and environment reviews of many others. The position of HAZOP study facilitator is nearly perfect for observation of the design process. The facilitator has to ask provoking questions, taken from a standard checklist, and to encourage the analysis process, but otherwise he or she should stand back and observe what happens. There is also a scribe to provide thorough recording of the process.

The most important method of design in process engineering is *design by copying*. Taking existing concepts and drawings and adapting them to a new application is a necessary process. Without it, it would be necessary to learn the old lessons repeatedly. Ideally, a new design would be an exact copy of an earlier successful design. However, the importance of adapting can be seen from the less successful places. It is not uncommon to see drawing errors repeated from plant to plant.

CASE HISTORY 16.2 Design Prehistory Can Still Be Traced

In one design, the need for a pipe connecting two trains of a plant extension was queried. It turned out that this had been required to allow gas filling on an earlier plant in which fuel gas had been needed. The feature had been carried forward through three design projects, unnecessarily.

After copying is done, it is important to ensure that the design follows any updated design philosophies. New requirements, introduced when setting general principles, are easily overlooked when a large set of drawings is adapted, especially if the designers are not familiar with the drawings.

CASE HISTORY 16.3 Copy by All Means, but Copy from the Right Place

A company decided to eliminate the use of corrosion coupons in the entire plant, because the coupons, used to measure corrosion rates, required opening the coupon mount, and several accidents had occurred with this. The coupons were replaced by improved ultrasonic testing. However, at the detailed design review for a new plant unit, the corrosion coupons were still found on the drawings.

When designing by copying it is also important to ensure that the correct design is copied.

CASE HISTORY 16.4 When Copying an Old Design, Make Sure That It Works

In a large refinery extension it was found that nearly 5000 people would be exposed to the risk of hydrogen sulphide release from a nearby sulphur recovery unit, which would be in operation throughout the construction period. It was decided to provide an alarm system and a workable evacuation plan. The decision was not entirely altruistic—in case of an accident the company involved could incur liabilities of several billion dollars. A concept for the alarm system was worked out and proposals were submitted to three design contractor companies. The first, quite honestly, said it could not make the design but would be happy to do the detailing and construction. The second produced a design very much in line with the concept and involving 90 detectors. The third company produced a design requiring 740 detectors. When asked what was the philosophy for the very expensive design, they said they did not know. The design was fast track and correspondingly had been copied from an earlier project. When asked to contact the designers, they said they had left the company, but they would contact the earlier client. After a 30-minute pause they returned to the proposal assessment meeting and asked if they could have an extension, to make a new proposal. It turned out that the earlier design was based on placing a detector at each flange and at each valve seal. It produced so many nuisance alarms that the client had switched the detection network off and had never operated it.

Design by algorithm is used for many designs and can be found in textbooks. The approach is typically used for unit processes, such as pumps and heat exchangers of distillation columns. The requirement is stated, and design alternatives are selected from a fixed list of possibilities. For a distillation column, for example, these might be the column diameter and height, the number of trays, the tray type, the level and flow of the reflux and the number and type of reboilers. For each alternative the performance of the equipment is calculated for a range of operating conditions. The design is gradually refined until the optimum is reached. If the designers are wise, they will then calculate whether the design still works under varying operating conditions. In particular they will calculate the possible turndown, that is, the lowest production rate at which the unit will still operate. They will also calculate whether the unit will operate well under hot standby conditions.

Algorithmic design will work well provided that details are taken care of as well. Lieberman, in *Process Design for Reliable Operation* [10], describes how an inexperienced designer, in the design of a distillation reboiler, forgot to take into account that outflow over the baffle of the reboiler would bring the liquid level higher than the top of the baffle. Outflow requires a certain depth of flow. This carried the liquid surface closer to the vapour outlet line at the top of the reboiler, so that there would be liquid carryover into the vapour return line. This prevented the reboiler from working, because the flow rate through the reboiler relies on the difference in density and pressure head between the liquid flow to the reboiler and the vapour return to the column.

If this sounds complicated, there is a reason. It is complicated. In Lieberman's book it took four pages and two diagrams to explain, in what is a very well-written book. It shows the level of physical visualisation needed for design.

Lieberman's book is subtitled *How to Bridge the Knowledge Gap between the Office and the Field.* Attention to practical details like this can make the difference between a successful design and a failure. And designers of large plants rarely have much experience of operations. This is especially true for design consultancies working in FEED. The engineers carrying out the design will usually have left the project for another before construction is even begun. The presence of experienced operators in HAZOP studies is therefore of enormous importance.

CASE HISTORY 16.5 A Conflict of Philosophies

In the design philosophy for a new plant, it was specified that all drain lines should have a block valve just after the vessel nozzle, then a spade (slip plate) providing an absolute shutoff, then another block valve. The reason for the spade was to prevent the release of oil into the drain system in the case of the valves passing, i.e. leaking internally. The motivation was environmental protection, limiting the amount of waste oil released to the drain and oily water–treatment system.

When the plant was to be drained, the pressure would first be reduced to atmospheric, then both block valves closed, the spade flanges unbolted and the spade replaced by a spacer (a ring of steel).

In the HAZOP study team was a young but very competent operation supervisor. He pointed out that the design would involve opening flanges on a section of pipe filled with oil. The oil would leak out and probably cover the maintenance fitter.

To give the design leader his due, he reacted very well. He asked for a 10-minute pause and went outside with his chief piping engineer. When he returned he just said, 'OK'. The earlier design was deemed to be an error and recorded in the problem action list. During the weekend, all P&ID draftsmen and draftswomen worked overtime and revised 220 P&IDs.

CASE HISTORY 16.6 Unknown Physics

In a course on process safety design in a fine chemicals company, the syllabus reached the subject of liquid hammer (see Chapter 7). The author asked how many of the engineers had heard of water hammer. Eighty percent raised their hands. When asked how many could calculate it, just twenty percent raised their hands. When asked how many knew about liquefied gas condensation hammer, just two persons raised their hands. Knowledge about condensation hammer is rare. The phenomenon involves liquid flowing into a pipe, condensing the gas in front of the liquid wave, the gas therefore presenting no resistance to flow and the liquid accelerating very rapidly. This effect caused rupture in a propane system at Texas City in 1984, and a major catastrophe ensued. The effect is not widely known except among boiler design engineers and boiler operators. When asked about how the two knew about gas condensation hammer, they replied, 'It happened to us last week. A 1-inch chlorine line ruptured'.

These case histories illustrate something about design error. Many errors arise due to lack of knowledge on the part of designers. This can be due to lack of knowledge about physical phenomena or lack of knowledge about details of how equipment works.

Brainstorming design involves most of the same steps as algorithmic design. However, the first step involves the use of free imagination to generate the design alternatives. There are several tricks for good brainstorming, including letting each participant in a brainstorming workshop generate ideas and write them down. This is followed by letting each person in the team present one idea at a time, so that all can contribute. Then everyone is allowed to elaborate on and suggest improvements on others' ideas. Generally, a filtering of ideas is necessary when idea generation has slowed down, to investigate advantages and disadvantages (pros and cons) of each design idea. Filtering saves a lot of effort in later analysis.

Brainstorming can create many original designs, but such designs can readily fail because they have not been tried in practice.

CASE HISTORY 16.7 Bureaucracy Can Defeat Any Design

A new oil-production system was to be developed in a large river estuary. No production plant was allowed onshore because the area was a natural park and nature reserve. Original development was made on a barge, but this involved several design problems, especially for escape if an accident occurred (there were both sharks and crocodiles in the river) and difficulties in making connections from pipelines to the barge in a location with high currents and tidal changes.

It was suggested in brainstorming that the development be made on barges, for ease of transport in placing the equipment, and then with piling isolating the barges and creating islands with backfill. The plant would effectively be onshore, and much safer, but it would not be in any prohibited location. After some effort in design, the concept failed. The design was deemed to be onshore, and company rules required sports facilities for onshore installations. Thinking outside the box does not necessarily mean that you can escape from the box.

Mental simulation does not create new designs but it does check concepts and generally adds details. Most mental simulation involves positing a change and then tracing its effects through the plant. For example, filling a distillation column could involve starting a pump, then opening the discharge valve. Liquid flows into the trays, then runs down into the bottom of the column. The level rises.

Most such mental simulation is expressed in terms of events, but some engineers can imagine continuous changes to some extent. These cannot be by their nature verbal, but at least limited parts of a plant can be pictured in the mind's eye, and linked changes in variables can be imagined. In tests of this for the column, for example, engineers pictured the liquid in the column increasing in temperature, the colour changing gradually from blue to red and the number of bubbles in the reboiler increasing. Then they imagined the reflux working and waterfalls of liquid flowing down the column. The imagery is obviously derived to some extent from the kinds of illustration used in training material and sometimes those on control systems displays.

Design detailing often involves placing symbols on drawings using standard patterns for each function. Such design is rarely done by a lead engineer. Errors can be made by misunderstanding the intent of the design and using the wrong symbol or by simple errors in naming and numbering or selection of symbols. Sometimes omissions are made because the drafters or detail designers do not have enough information. For example, a valve may be marked *LO* for 'lock open' or *LC* for 'lock closed', but if the operator does not know the meaning of each mark, the indication may be omitted from the drawing, 'to be included later'. This leaves ample scope for errors.

DESIGN ERRORS DUE TO LACK OF KNOWLEDGE

The description of types of design given above indicates some of the kinds of errors which can occur. This chapter gives more examples. A general list of error causes is given in Table 16.1. It is based in all on 250 abnormal occurrence reports from nuclear reactor operation, 35% of which involved design error [1]; 121 accident reports in the book on chemical industry accidents by Drogaris [11], 55% of which involved design error; several hundred accidents described in the U.S. RMP database 5-year accident histories and an in-depth analysis of some 2000 design errors observed by the author from plant safety audits and HAZOP studies.

This chapter gives several causes or causal influences which lead to design error. Many are illustrated by items found in HAZOP studies, but all are also exemplified in the collection of design-related accidents made as part of the studies for this book.

It can be seen that lack of knowledge is a critical contributor to design error.

CASE HISTORY 16.8 A Grievous Lack of Knowledge

In a refinery, operating pressure was specified to be equal to design pressure on hydraulic tubing used for valve control. This is a very bad error, because there will always be disturbances which push the process variables above the normal operating point. There is still a safety margin between the design pressure and the actual rupture pressure, but this is intended to look after variabilities in materials and in construction, not to serve as a margin for operation.

In the event, the tubing pulled out of a compression nipple when disturbed by an operator. The pipe whipped, slicing into the operator's helmet and cutting the top of the helmet away. The operator suffered neck injury.

TABLE 16.1
Design Error Causes for Process Plants

Error Type/Cause	Description	Examples
Previously unknown phenomena	The accident occurs due to a physical or chemical phenomenon which is unknown at the time of design.	• Gas condensation hammer at Texas City [12]. • Reaction between rainwater and zeolite with absorbed H_2S [13].
Lack of physical or chemical knowledge	The designer is unaware of a physical or chemical effect.	Sodium hypochlorite decomposition to chlorate, with chlorine release [14].
Lack of detail knowledge	The designer is unaware of equipment details.	No explosion panels were fitted to a reciprocating compressor crankcase because the designer had not known about crankcase explosions.
Lack of operation knowledge	The designer is unaware of physical effects occurring in a real plant that are not available in references.	Fatigue cracking and rupture in a high-pressure rain line due to looseness in a screw jack pipe support arising from vibration.
Lack of knowledge of standards, guidelines or project philosophy	The necessary requirements are not known to the designer, who does not have access to the documents or has not read them.	Design of a gas-processing plant according to the older API 521 standard for relief systems, which describes dimensioning relief systems for pool fires (which could not occur on the plant) and not for jet fires.
Wrong material for construction	The design uses a material which is unsuitable for the application. This may be due to lack of knowledge on the part of the designer.	Lack of knowledge of the inappropriateness of low-silicon carbon steel for sour service (with H_2S in the oil) leads to sulphidation thinning, rupture and an unconfined vapour cloud explosion.
Lack of information on the purpose or requirement of the design	The design team leader is unaware of the needs of the team because of oversight or egoistic reasons, such as the desire to protect one's own job.	• Wrong philosophy for drain system isolation due to a requirement which was communicated to the designer only in the final HAZOP workshop. • The control system designer for a design company was the only person knowing the specifications for the power-cogeneration systems. Although overworked, he did not share necessary details with any others.

(*Continued*)

TABLE 16.1 (CONTINUED)
Design Error Causes for Process Plants

Error Type/Cause	Description	Examples
Simple clerical errors	Typing mistakes and similar problems occur.	Primo Levi describes a difficulty in making a paint process work because of a speck of fly dirt on a manufacturing recipe [15].
Oversimplicity in mental modelling	The designer models the process mentally but cannot capture all the details of the internal working of the plant.	Operators did not anticipate 'swell' due to bubble formation in liquid reactants in a batch reactor, resulting in overflow of the liquid into the vapour line. The reactants froze, so that there was no venting of the reaction vapour. The vessel overpressured, the disk burst and flammable solvent vapour was released. The designer had assumed that there was plenty of room in the reactor for the reaction to occur and the vapour to disengage from the liquid.
Limited scope of mental modelling	The designer thinks of just some part of the process (the most critical part or the part he or she knows about).	Most designers do not consider the problem of pressurised draining in their designs, even though this can lead to accidents and gives a significant number of accidents (about 0.5% of all process plant accidents).
Inaccurate mental simulation	The designer imagines the sequence of events or processes incorrectly.	In adding a new process unit to a plant, feed was taken from a common inlet line. Unfortunately, the flow was two phase, i.e. liquid and gas. When built, most of the gas went to one processing train, most of the liquid to the other, so that both trains were overloaded.
Limited mental simulation of normal operation only	The designer does not design for plant start-up and shutdown or for limited operation such as hot standby.	In an alcohol distillation unit, the designer did not allow for draining the plant prior to maintenance. There were no drains for several low points and no connections to allow for water washing.

(*Continued*)

TABLE 16.1 (CONTINUED)
Design Error Causes for Process Plants

Error Type/Cause	Description	Examples
Nonanticipation of component failure in the design	The designer plans neither for component failure, misuse or overloads nor for external effects.	The designer interconnects several tanks of different heights, allowing liquid to flow from one to another if valves are not lined up properly, thus allowing overflow to occur.
Design working on out-of-date releases of drawings and specifications	Various disciplines in the design work in parallel. Whenever a stage in design is complete, it is issued as a new revision. However, in the meantime some disciplines are working on out-of-date design material. If the different parts of the design are not brought fully into alignment, errors occur.	The control system on one design did not take into account a cross coupling between two production trains, introduced late in the design in order to allow for partial shutdown of each train while retaining near-full production.
Relying on out-of-date drawings when modifying plants	The designer makes use of the drawings which are available and does not make a survey of the actual plant when planning an upgrade or extension.	On an upgrade which would allow new pumps to be added to increase capacity, a vent line which had been added was overlooked, so the new design drawings did not include the vent line. When the plant was started during commissioning, it proved impossible to get the pump to work.
Last-minute modifications made during commissioning	Plant commissioning is a difficult time, and any errors may require improvisation. The modifications are not always checked for safety.	Cryogenic ethylene pumps needed to be cooled down slowly. Vent pipes were provided for this so that small amounts of liquid could be admitted and allowed to evaporate. Ice formed in the vents when it rained, so the vent discharge was modified so that it pointed downwards towards the ground. This created a hazardous vapour cloud at every start-up.

PREVENTING DESIGN ERROR

The most important method for preventing design error is to have good design standards and people who know how to use them. Real knowledge, though, comes from making a design and following it through from sketch to commissioning and operation. This becomes less possible with modern design practice. The industry is divided and engineers in design have often left a project before commissioning begins, while operations engineers only rarely transfer into design.

A HAZOP study is a procedure in which disturbances in process variables are proposed and possible causes and consequence of the disturbances are considered and tabulated. Then the existing safeguards are recorded, and if these are insufficient, additional safeguards are proposed or ways of eliminating hazards are investigated. The disturbances are identified by means of physics-based checklists, so that it is possible to get a reasonably complete starting point for the analyses.

The most important aspect of a HAZOP study is that it is, for the most part, carried out in multidisciplinary workshops. These workshops are one of the few places where designers from different disciplines are forced to work actively together for long periods, discussing designs in depth. The quality of a HAZOP study improves very much if experienced operators are included in the HAZOP team, and this has become standard practice.

From its early days in the mid-1970s, where many companies regarded a HAZOP study as an unnecessary academic exercise, HAZOP studies have developed to become a legal requirement, in the U.S. Occupational Safety and Health Administration Code of Federal Regulations Title 29 Part 1910, and, by now, to an essential part of the design process. In a large project, HAZOP studies may be performed as many as four times, at different stages of the design, and there is in many countries a requirement for periodic update, typically every 5 years.

The HAZOP process was validated by the present author and colleagues in the 1970s [16], by repeating analyses with independent teams and by actually building and operating the plant and following its performance for several years. The method was also validated by Suokas [17]. Since then a continuous check has been maintained on the quality of the HAZID [18]. Considering the trust placed in HAZOP studies, the actual performance is rather disappointing. The degree of completeness of accident cause tabulations is typically about 85% and rarely exceeds 95%. Even when corrected for importance (relative risk) of different causes, the completeness is rarely above the same 95%.

Reasons for incompleteness of the hazard and design error identification, derived from a study of 30 HAZOPs, are in Table 16.2. The errors were identified by using an automated version of a HAZOP study [19] and by detailed follow-up.

Because of the problems of incompleteness in analyses, companies have turned to lessons learned from earlier accidents as a supplement. Many companies now have documented their own lessons learned. There are also searchable collections of data. The Institution of Chemical Engineers publishes *Loss Prevention Bulletin*, which is published six times yearly, with often 10 accident reports. These are commented so that maximum information and relevance can be obtained. The Center for

TABLE 16.2
Causes of Incompleteness in Analyses (Observed in 30 HAZOPs with Follow-Up)

Cause of Incompleteness	%	Possible Improvements
Lack of knowledge		Experiments
• Of the whole scientific and engineering community	0.2	Pilot plants
• Of the individual	6	Literature Engineer studies Case story collections Training in risk analysis
Too narrow a framework for the analysis	1–5	Guidelines for scoping analyses
Analysis of documents which do not reflect the system	7	On-site inspection As-built drawings
Errors in drawings	1	As-built drawings used for HAZOP study
Limited resources for analysis	22	
Simple oversights	5	High-quality checklists
Masking (one error hiding behind another, so that all effort goes into identifying the first error)	2	Improvement in analysis procedure
Judgement errors (hazard is identified but considered unimportant, sometimes due to misleading experience, sometimes due to lack of knowledge about the actual frequency)	5–15	Quick rule-of-thumb calculations which can be applied in the HAZOP study Good access to design calculation tools A long list of accident cases
Problems introduced later	4	Follow-up of actions Change control

Chemical Process Safety (CCPS) has developed a database for accident case histories and lessons learned, which is, unfortunately, available only to CCPS members. There are several books giving this kind of information, including a long series by Trevor Kletz [19].

One of the problems with these collections of accident case histories is that they need to be read and remembered or they need to be searched, which takes time. Systematic lessons learned analysis was developed so that relevant accident case histories can be looked up as HAZOP results are being recorded, with a single button click retrieving relevant case histories during consideration of each disturbance/deviation. It has allowed HAZOP study completeness to be enhanced to above 99% when comparing independently prepared HAZOP studies.

In view of the amount of effort which goes into a HAZOP study, it is surprising how little information from it is communicated to operators. Even when operators take part in HAZOP studies, there will only be one or two persons; most operating staff will not be involved. When the HAZOP study is completed, it is usually archived permanently, and only the recommendations go forward to the design team.

CASE HISTORY 16.9 Using Hazard-Analysis Results

An in-depth HAZOP and risk-reduction project was carried out for a paint factory which had experienced several serious incidents. After completion of the project, a table of incident causes, the symptoms as they would appear to the operator and the required operator responses were described in a handbook. The operators became very angry. Their comment was, 'Why wasn't this done 15 years ago?'

Many design errors are not visible from the P&IDs. These include dimensioning errors, piping design errors and buildability errors, such as inadequate access, interferences between piping and vessels and lack of escape routes.

DESIGN REVIEW

Design review is a key element in finding errors. The term covers a wide range of methods including the following:

- Recognising patterns in P&IDs which are typically sources of problems, such as changing the pressure specification on the wrong side of a block valve (this results in the valve being selected for the lower-pressure side, rather than for the higher-pressure side, and possibly bursting under high pressure)
- Inspecting drawings for items which 'look wrong', based on general experience; often this is just a search for novel configurations which may be novel because they are wrong
- Using design rules of thumb to check dimensioning of pipes and vessels

The success of HAZOP workshops has led to several other workshop types to be introduced into the design process. These include the following:

- Inherent safety reviews [20].
- 3D model reviews, involving looking at 3D plant models and considering access, interference, space for materials transport, space for maintenance, crane hardstanding for lifts, safe lift paths and access for equipment installation.
- Human factor workshops.

CAD has introduced many advantages in checking designs, such as in dimensioning of vessels, tanks and piping. The use of dynamic process simulation has helped especially in reducing many kinds of errors, and algorithms for dimensioning of equipment are widely available. Unfortunately, there is a tendency for less experienced engineers to regard the output from computer programs as absolute truth while at the same time losing many manual design skills. Errors in software can then result in catastrophe.

CASE HISTORY 16.10 Underdimensioning Leads to Platform Collapse

Concrete pedestals for an oil-production platform were designed using finite element analysis. The software contained an error. As a result, the walls of the pedestal were too thin. This was noted by workers carrying out the construction, but the work proceeded.

When installing the topsides (upper part) for the platform, the pedestals were partly filled with water, so that the structure lowered in the water. The topsides were then to be sailed over the pedestal. However, the water pressure on the concrete was too great. The pedestals imploded, and the entire structure sank. This delayed development of the intended gas field by over a year.

Software errors are increasingly becoming a problem in design. One of the main problems is that many programs are 'black boxes'. It is not possible to see what the program is doing. As a result, the users are placed in a difficult position—errors cannot be observed and suspicious results cannot be investigated. Also, since much of the software is made by programmers rather than engineers, it is often difficult to obtain adequate explanations in cases of doubt.

CASE HISTORY 16.11 Beware of Software when Designing

A program was being used to calculate the effect of jet fire on a control room building. The jet fire could arise from a break in a pipeline carrying a mixture of oil and natural gas. When calculated, the jet fire and radiation appeared small. It turned out that the program did not have a capability for calculating liquid jet fires, so that part of the calculation was ignored. The program calculated for gas and liquid separately. Once this was explained, a 'work-around' was found in which the liquid was represented as a very heavy gas. A more reasonable result was obtained, but by this time the protection for the control room was in question.

As a result of problems of this kind, best practice requires calculations to be made by two different sets of software, ideally not using the same models or formulas. Computer fluid dynamics is often used to spot check calculations in this way. Alternatively, rule-of-thumb calculations can be made to check that software results are at least within a reasonable range of the correct result.

REFERENCES

1. J. R. Taylor, Design Error in Nuclear Power Plants, Risø National Laboratory M-1724, 1974.
2. J. R. Taylor, A Study of Failure Causes Based on U.S. Power Reactor Abnormal Occurrence Reports, *Proceedings of the Symposium on Nuclear Safety*, Vienna: International Atomic Energy Agency, 1975.

3. P. Haastrup, *Design Error in the Chemical Industry*, Risø National Laboratory Report R-500, Doctoral Dissertation, Technical University of Denmark, 1984.
4. J. R. Taylor, Understanding and Combating Design Error in Process Plant Design, *Safety Science*, Vol. 45, Nos. 1–2, pp. 75–105, 2007.
5. J. R. Taylor, Statistics of Design Error in the Process Industries, *Safety Science*, Vol. 45, Nos. 1–2, pp. 61–73, 2007.
6. K. Kidam and M. Hurme, Origin of Equipment Design Errors. *Journal of Loss Prevention in the Process Industries*, Vol. 25, Issue 6, pp. 937–949, 2012.
7. Failure Knowledge Database, http://www.sozogaku.com/fkd/en/, Available online 29 May 2011.
8. IEC 61882, *Hazard and Operability Studies*, IEC, 2001.
9. ISO 17776:2000, *En Petroleum and Natural Gas Industries – Offshore Production Installations – Guide to Tools and Techniques for Hazard Identification*, Geneva, Switzerland: International Standards Organisation, 2002.
10. N. P. Lieberman, *Process Design for Reliable Operations*, Gulf Publishing Company, 1983.
11. G. Drogaris, *Major Accident Reporting System: Lessons Learned from Accidents Notified*, Amsterdam: Elsevier, 1993.
12. J. A. Davenport, Hazards and Protection of Pressure Storage of Liquefied Gases. In *5th International Symposium on Loss Prevention and Safety Promotion*, European Federation for Chemical Engineering, 1986.
13. Anon, Failure to Recognise a Hydrogen Sulphide Hazard, *Loss Prevention Bulletin* 215, October 2010.
14. P. Levi, The Periodic Table, 1975.
15. C. Sandom and R. S. Harvey, ed., *Human Factors for Engineers*, London: The Institution of Engineering and Technology, 2004.
16. J. R. Taylor, O. Hansen, C. Jensen, O. F. Jacobsen, M. Justesen, and S. Kjærgaard, *Risk Analysis of a Distillation Unit*, Risø National Laboratory Report M-2319, 1982.
17. J. Suokas, *On the Reliability and Validity of Safety Analysis*, Espoo, Finland: Technical Research Centre of Finland, 1985.
18. J. R. Taylor, *Forty Years of HAZOP – Lessons Learned Loss Prevention Bulletin.* Rugby: Institution of Chemical Engineers, October 2012.
19. T. A. Kletz, *What Went Wrong*, Fifth Edition. Elsevier, 2009.
20. T. A. Kletz, *Cheaper, Safer Plants or Wealth and Safety at Work – Notes on Inherently Safer and Simpler Plants*, Rugby: Institution of Chemical Engineers, 1984.

17 Errors in Procedures

In nuclear power plants in the late 1960s and early 1970s, errors in operating procedures were found to be nearly as large a source of accidents and incidents (10%) as were operator errors (12%) (see Chapter 1) (Table 17.1). For oil, gas and chemical industry plants, similar results were observed in reports of major accident hazards [1].

The frequency of errors in written procedures depends, of course, on the quality of work that goes into them. Various approaches were observed in new plants:

- Preparation by experienced operation engineers
- Preparation by design engineers
- Preparation outsourced to low-cost contractors, with checking by experienced operators

Of these, only the first was found to provide good results from the start, but good procedures generally involve considerable review by experienced operators.

WRITING OPERATING PROCEDURES

Operating procedures can be in several styles with different scopes. There is a need for simple checklists, so that the sequence of operations can be maintained and nothing overlooked. There is a need for straightforward procedures to be used as a reference in the control room. And there is a need for tutorial-type procedures, to be used for training.

A typical organisation for modern procedures includes the following sections:

- Purpose of the unit
- Design basis, operating data, mass balance product streams, chemicals and handling
- Process description
- Process control, control narrative, control loops and complex controls
- Quality control (QC)
- Safeguarding systems, alarms and trips, pressure relief and depressurisation
- Interconnections with other units
- Operating envelope and matrix of permitted operations
- Initial start-up procedures
- Normal start-up and operation
- Normal shutdown
- Upset, disturbance response, shutdown and depressuring
- Instrument summaries

TABLE 17.1
Incidents and Accidents due to Errors in Procedures in Nuclear Reactors in the 1960s and Early 1970s based on 44 Incidents

Error Type	%
Omission of a step in a procedure	44
Omission of a complete subprocedure	12
Omission because a physical effect or equipment peculiarity was unknown when the procedure was written (new knowledge)	16
Procedure ambiguous, open to misinterpretation	7
Wrong test frequency specified	2
Entirely wrong procedure specified	2
Extra controls or checks required	6
Error with unknown cause	14

- Appendices—process flow diagrams, Piping and Instrumentation Diagrams (P&IDs), mass and energy balances, safeguarding of flow diagram, cause-and-effect matrix, material safety data sheets and equipment specification sheets

It can be seen that a large part of this material is pedagogic. Trainee operators will study this and take an examination before receiving an operator's certificate. The examination also includes tests of ability to apply the acquired knowledge.

In the actual example from which this list of contents was taken, the style is purely technical. There is little explanation of background or explanation of why particular steps are taken. In the actual example, there were no cautions or warning concerning potential hazards except in the section concerning handling of chemicals. Contrast this with military training manuals, which are typically well illustrated by photographs and do include information directly intended to ease learning. Some of the companies audited do have excellent tutorial material in their training departments, apart from the official operating manual.

The quality of operations and maintenance procedures varies widely, even in well-managed plants. Procedures vary from nonexistent to poor and incorrect to a good standard. It is quite rare, though, in any plant, to find procedures which are referred to by operators except check lists for infrequently performed procedures. They tend in practice to rely on their own experience and on learning from others. The procedures are used primarily in training, before operation begins or as references for looking up operating data or drawings.

The reason that operating procedures are so rarely used is that the style is difficult to understand and is often inadequate.

Computer programs are available today to support all these styles of procedure. Programs can generate each style from a single database. Procedures can also be stored on a computer network so that they are more easily available.

MAINTENANCE MANUALS

It is good practice to have a full set of maintenance and safety procedures. For the best plants, a complete set of such procedures often fills many shelves of ring binders. Some of the topics to be covered are:

- The list of instruments and apparatus to be routinely tested
- The procedures for testing
- Routine QC procedures for materials and products
- Labelling requirements
- Getting ready for repairs
- Shutoff of electrical equipment
- Hot work procedures and permits
- Periodic inspection of vessels and corrosion
- Routine apparatus maintenance procedures
- Safe methods of sample taking
- Tank level measurement procedures
- Use of cranes
- Use of breathing apparatus and gas masks
- Safety clothing
- Filling and emptying of tanks for hazardous materials
- Renewal of hoses
- Renewal of crane slings
- Procedures for handling of substances
- Safe storage and warehousing procedures
- Gasket and seal replacement
- Special precautions in selection of materials for repairs
- Repair and modification authorisation procedures
- Pipe deblocking procedures
- Procedures for safe use of compressed air equipment
- Marking and clearing of escape routes
- Carrying flammables, toxics, etc.
- First-aid treatment of burns, acid burns and poisoning
- Testing of fire-protection equipment and safety showers

Most of these procedures are stored on the intranet, many of them inside large maintenance scheduling and reporting databases such as SAP or Maximo. Most of the detailed maintenance procedures for specific equipments are generally written by the vendor companies. In audits, such manuals were found to vary enormously in quality, from well written and extensive (~150 pages) for a single instrument to superficial and poorly written for a large compressor. Quite obviously, the company supplying the compressor did not intend the purchasers to perform their own maintenance.

METHOD STATEMENTS

For virtually all maintenance tasks, method statements will be prepared. These state in detail the following:

- The purpose, nature and scope of the work to be done
- The preconditions and starting requirements for the task
- The work permits required
- Individual responsibilities
- Step-by-step procedures
- Tools and equipment required
- Any heavy lift assessments required
- Safety requirements (equipment and staffing)
- Safety instructions

Such method statements will generally be reviewed and signed by maintenance engineers. Preparation is time consuming, so generally the statements are archived and are written generically so that they can be reused, with modification.

Method statements are generally subjected to job safety analysis (JSA) or task risk assessment (TRA). These are generally carried out by maintenance teams in workshop style, often with the presence of operators. The assessments provide additional recommendations for safety measures. The TRAs or JSAs are generally subject to approval by safety engineers. The TRAs are also time consuming to complete, so these are generally made generically. When a new task of a given type is to be made, the standard TRA is taken and reviewed for applicability. It is also used as a checklist to ensure that all safety measures recommended in the TRA are in place.

In audits it was found more or less consistently that TRAs were *very* generic. As risk analyses they would generally be regarded as 'better than nothing, but not good'. This applies for in all 12 companies where audits were made of the TRAs. Senior managers regularly complained of cut-and-paste approach to risk analysis. The main reason seems to be the lack of validated methodology and of quality standards. An example of an approach which has produced good TRAs, which could be validated, is given in Chapter 20.

ERRORS IN PROCEDURES

Errors in procedures were investigated as part of the safety and QC reviews for plants preparing for commissioning and built since 2005. Note that reviewing procedure during commissioning shows procedures at their worst. Commissioning engineers often perform red marking of procedure manuals for update before plant handover to the client and in this way provide the final QC, as well as updates to take into account the changes made.

By far the largest group of errors found in the procedures were simple omissions. Many are trivial, but these arise because focus is the main strand of the narrative, that is, the main flows of materials, not the supporting systems. A typical example is forgetting to start the cooling water in a distillation unit.

The second most frequent error was in referring to equipment tag numbers and omitting or getting wrong the equipment functional name.

A major error is the omission of cautions and warnings from the procedures. In the way procedures are written, there is little in the way of safety assessment, so that naturally there is no safety information in the procedure. A typical example is the warning 'Raise the temperature gradually, no more than 2° in 5 minutes, in order to avoid stress cracking in the vessel wall'.

Developing a full set of cautions and warnings is difficult and really requires an analysis like that described in Chapter 18.

Another error type is describing a procedure which will not work. For example, a procedure for starting a pump was described in which a second pump was to be started while its twin was still in operation. When the procedure was carried out, both pumps inevitably tripped because of low pressure in the suction lines.

CHECKING WRITTEN PROCEDURES

Written procedures can be checked by step-by-step review, marking up the status of plant equipment and fluid flows on process flow diagrams or piping and instrumentation diagrams at each step.

This approach was automated using a simulator approach originally developed for automated HAZOP analysis [2]. A model of the plant is obtained by inputting P&IDs and cause-and-effect matrices for the actual plant and the interpreting the drawing by pattern recognition. The control interface is input in the form of photographs, on which actuation is marked. Then the model can be 'operated' by manipulating the virtual controls. The effects of operation are displayed on the drawings.

The approach allows a number of checks to be made which are difficult or time consuming for manual assessment, including presenting equipment and action-specific checklists and retrieving lessons learned from earlier difficulties.

PROCEDURAL DRIFT

When a plant is first built, operating procedures are written to allow the plant to be operated as intended. The initial procedures always contain a few errors and are generally marked up for correction during the commissioning period.

Almost as soon as the procedures are completed, they begin to be out of date. The plant is modified slightly, the company works hard during the first years to increase output and operators 'improve' procedures to make work easier. A classic example is a change in a procedure when an operator is required to access a valve that is inaccessible or requires a long climb.

Even with very well-regulated procedures, there will be changes. Change in raw material quality or change in product specifications forces change in operating parameters and sometimes requires change in procedures too. An example in oil fields and processing plants of such change arises due to a 'water cut', which results in incoming oil increases. Operators may then have to drain off water from various parts of the plant, and changes are made as the problem worsens.

Written procedures may or may not be updated for these reasons, and procedural drift may occur, where the actual procedure differs from the written procedure.

Procedural drift becomes insidious if they improve production rate or improve convenience but at the same time push the plant to the limit of the operating envelope.

CASE HISTORY 17.1 Procedural Drift [3]

Hydrocrackers are intended to take heavy oil compounds and hydrogen and turn them into lighter molecules, suitable as high-value fuels such as gasoline and LPG. They do this at high temperature and high pressure, with a catalyst bed to guide the molecular process.

In the actual accident case, heavy oils were treated with hydrogen at high temperature and pressure through a catalyst bed that turned sulphur in the oil to hydrogen sulphide and nitrogen to ammonia. These substances would otherwise poison the stage 2 catalyst.

In stage 2, hydrogen was again used in a high-temperature and high-pressure reactor, to break up large molecules and produce propane, butane, diesel oil and heavier fractions. Hydrogen from another plant unit was fed to, and recycled around, each stage. The pressure in the second stage was about 80 bars; the temperature varied through multiple catalyst beds (five in all for each of three reactors). The maximum reactor temperature alarms were set at 800°F, above which reaction runaway could occur, with overheating and possible vessel failure. The maximum outlet temperature was set at 690°F, after which downstream equipment damage could occur.

Monitoring any reactor with solid catalyst beds is a problem if the reaction is exothermic (produces heat). Hot spots form if the flow through the bed is not uniform and well distributed. Hot spots which are hot enough to soften the catalyst or to turn the oil to coke will block part of the flow. This can cause higher local flow rates or reduced flow rates in other parts of the reactor bed and can reduce cooling flows.

For these reasons, the reactor beds for each reactor were instrumented within all 96 thermocouples. Forty of these measurements were relayed to the control room, and 56 were displayed on a field displayed beneath the reactor. Forty thermocouples were connected to alarms in the control room. There was also a data logger for the temperatures with alarms for temperatures more than 50°F above or below the average bed temperature.

There had been a good deal of trouble with a guard reactor which was used to turn nitrogen into ammonia and sulphur into hydrogen sulphide. This was needed to protect the reactor catalysts. On the day of the accident, the feed rate was reduced in order to allow the temperature to increase, to compensate for the poor conversion of nitrogen. There was also a leak on a cooling heat exchanger, so this was closed down.

The temperature in the hydrocracker reactor rose. Hydrogen was admitted in order to provide a temperature quench. However, at higher temperatures, the hydrogen can react with the naphtha, forming methane. This releases heat and causes the temperature to rise. The operators struggled to keep the hydrocracker running for about a day and a half and for 7 minutes to control temperatures. After this, the exit temperature from the reactor was too high. The exit piping yielded, then ruptured. Naphtha and hydrogen escaped. They ignited and exploded.

In the following investigation, pillars of coke were found in bed 4 of all three stage 2 reactors. Coking on the catalyst necessitates higher operating temperatures to achieve good hydrocracker performance. It was also found that after a certain stage it would have been impossible to cool the reactor bed sufficiently to prevent runaway.

The accident shows the reality of control for complex equipment, with many interacting parts. In the first event sequence, a mechanical failure (a leak) in one production train required shutdown, which, in turn, required redistribution of load to two other trains. This increased load poisoning of the catalyst in the reactors. Then with the reactor on high load and with uneven flow due to coking, runaway occurred.

There was a detailed emergency procedure for this case, which required the following:

- Change control set points to control the temperature by quenching with hydrogen.
- Sound an alarm, stop oil feed, reduce furnace firing and set hydrogen circulation at maximum at 25°F above normal.
- *For any temperature 50°F above normal, trip and depressurise the reactors.*

The procedure was posted on the operating board. The operators did not put the emergency procedure into effect. The accident investigators noted the following factors which contributed to this omission:

- Confusing temperature reading: The fluctuations led the operators to believe that the readings were erroneous. The data loggers returned to a default reading of 0 when the temperature exceeded 1400°F but this was not understood.
- There had been earlier failures on the data logger, including one a day earlier.
- A change in temperature-monitoring system had been made 10 days earlier but was removed again because of errors in calculations for unconnected thermocouples.
- The strip chart display for bed 4 temperature showed as normal, while the data logger showed alarm because the strip chart and the logger were connected to different thermocouples.
- Operators did not have access to all the thermocouple readings.
- *The operators had earlier successfully controlled temperature excursions by increasing quench, reducing inlet temperatures and stopping feed flow.*

- Earlier use of the depressurisation system had led to liquid release to the flare and in one case a grass fire and in another a flammable vapour cloud, so the operators did not want to shut down.
- *Supervisors and senior operators had not ensured that board operators shut down and depressurised at the required temperature excess on earlier occasions.*

The key issue was that the operators had become used to operating the plant to compensate for any low production; they had operated close to the safe limits and had become accustomed to ignoring the emergency procedure because they had found a 'better' way of controlling temperature.

Accidents with this kind of effect are not common in accident reports, but struggling with control is a periodic problem in some plants. Many incidents arise because new phenomena which were not considered in the original design occur or develop with time.

REDUCING PROCEDURAL ERROR

Apart from checking the completeness of the procedures, at least against the list in at the beginning of this chapter, checking using the AEA procedure in Chapter 18 is a very good idea. It will capture most mistakes, especially if it is carried out as a workshop which includes operators, and following HAZOP practice in which everyone has access to a set of plant drawings.

A good check list for writing procedures is given in Ref. [4]. The book by Sutton [5] gives a good tutorial introduction to writing procedures.

REFERENCES

1. J. R. Taylor, A Study of Failure Causes Based on US Power Reactor Abnormal Occurrence Reports. *Proceedings of the Reliability of Nuclear Plants International Atomic Energy Agency*, Vienna, pp. 119–130, 1975.
2. J. R. Taylor, J. V. Olsen, A Comparison of Automatic Fault Tree construction with Manual HAZOP, 4th Int. SympLoss Prevention and Safety Promotion in the Process Industries, I Chem E Symposium Series No. 80, 1983.
3. U.S. Environmental Protection Agency, *Chemical Accident Investigation Report: Tosco Avon Refinery, Martinez, California*, 1998.
4. ASM, Effective Procedural Practices, ASM Consortium. www.asmconsortium.org, 2010.
5. I. S. Sutton, Writing Operating Procedures for Process Plants, Southwestern Books; 2nd edition, 1995.

18 Identifying Human Error Potential

Analysis of operator error has developed gradually over some 40 years, starting with basic ergonomic studies during the 1950s, aimed at determining just what kind of errors operators could make. A tradition grew up around the study of aircraft control which involves determining the limitations of persons in directly controlling a system. This work tends to be mathematical and experimental.

One of the first collections of human error probability was that of Munger et al. [1]. Swain and Guttman, in the 1960s and early 1970s, developed a model of human error which included the concepts of error-prone situations [2], error recovery and performance-shaping factors.

Swain and Guttman also developed the Technique for Human Error Rate Prediction (THERP) method for predicting human error frequency. This involves identifying a range of 20 different types of error modes, i.e. the external forms of the errors, and a range of performance-shaping factors which would influence the error probabilities by multiplying tabulated baseline error probabilities by values of the performance-shaping factors and by error-recovery factors. The values of probabilities in this method were based on detailed observation, especially observation of various production lines. This method is still one of the most successful in predicting human error probabilities (see Ref. [3]).

Swain and Guttman's work is used today mostly for the THERP method which is still widely used in risk assessment. In fact, the background information on human performance in production and operations is, in my opinion, even more valuable, in that it provides concepts and illustrations which go well beyond the THERP method.

A more recent and more comprehensive approach to prediction which is similar in formulation to that of THERP is Jeremy Williams's Human Error Assessment and Reduction Technique (HEART) [4].

Both THERP and HEART are described in detail, with examples and validation studies, by Kirwan in Ref. [3]. Kirwan also provides a table of human error probabilities based on observations in nuclear plant and in simulator experiments.

The origin of the method described below was a project aimed at identifying all the potentials for human errors in order to design an interlock system after an explosion in a sodium methylate plant [5] in an effort to develop a method which would be practical, yet still made use of the ideas in the SRK model [6]. The method was validated qualitatively in a major comparative study in 1978 to 1980 [7]. It has taken a further 30 years to amass sufficient data to allow a quantitative validation.

The qualitative validation of the action error method was carried out by applying it, along with the HAZOP study, fault tree analysis and precommissioning audit methods, to a urethane/methane/water distillation unit, then building and operating the plant, with a period of tracking of 2 years. In all, three incidents and near misses occurred in the operation period, caused by a design error, an analysis error and an operator error. All three errors are interesting from the point of view of human error and are described in the report.

ACTION ERROR ANALYSIS

The problems with any human reliability–analysis method are, firstly, getting a systematic grip on where errors can arise and then dealing with the very large number of error possibilities which can be identified.

As a starting point, the action error–analysis (AEA) method requires either written operating procedure or a task analysis with an equivalent degree of detail. Task-analysis procedures are described by Kirwan [8], by Annett and Duncan [9], by Stanton [10] and by Embrey [11]. For most oil and gas plants though, there will be written operating procedures, and for reasons of credibility of the analysis, these should preferably be used as a basis, after marking up to incorporate any updates. This also has the advantage that errors in written procedures can be found.

If the plant to be analysed exists, the operating procedure should be supplemented by an observation of the actual performance of the procedure. In many cases it will be found that the actual procedure deviates from the written procedure. In many cases, the deviation from the written procedure occurs because it is physically impossible for the written procedure to be carried out. This, in turn, arises because the engineers who write the procedures do not know enough about the plant.

The first step in the AEA method is to identify the modes of error, i.e. the immediate physical effect of the error on the plant. The reason for using this starting point is that there may be many possible error causes, but there are only a few error modes possible at each step. By starting with the error modes, the problem may be simplified from the start because there may be only a few modes which are physically possible, and there will generally be only some of these which have a significant consequence.

For each step in the operating procedure, the following question is asked: 'What can physically go wrong with this step?' This question has little to do with human error and a lot to do with the physical aspects of the operation. For example, with a valve, you can omit opening it or open it too much, too little, too fast or too slow. These are the error modes. A list of key phrases for identifying error modes [6] follows:

1. Omission
2. Too early/too late
3. Too fast/too slow
4. Too much/too little
5. Too hard/too slight
6. In the wrong direction
7. In the wrong sequence
8. Repetition
9. Wrong object
10. Wrong substance

11. Wrong materials
12. Wrong tool
13. Wrong value
14. Extraneous action (one unrelated to the task but interfering with it)
15. Wrong action

Error modes 1 to 8 form a closed set, and a complete analysis can be made. Modes 9 to 13 require some imagination or close study of the actual site. Modes 14 and 15 require a lot of imagination or a good list of examples. Many of the key words will represent physically impossible or irrelevant errors for some kinds of action. For example, it is impossible to press a push button in the wrong direction!

Examples of extraneous actions are cleaning the control panel while operating (risking pushing a button at the wrong time); rebooting a control computer during start-up of a plant (presumably a redundant computer); drinking coffee while operating a keyboard and switching pumps to allow for periodic maintenance while someone else is bringing up the production level on a distillation column.

LATENT HAZARD ANALYSIS

At this point in the procedure, it is convenient to extend the definition of operator error to include performing the correct operations but in the presence of a latent hazard. For example, an air hose could be pressurised while there is an open valve connected at the end. The hose would then whip from side to side and could kill.

Latent hazards are one of the most frequent causes of accidents, being either the sole cause or a contributing cause, in some 20% to 30% of major hazard accidents in process plants.

There is a simple and efficient procedure to check for latent hazards:

1. For each operation, list each item of equipment affected by that operation as a sequence of events. For example, trace the effect of opening a valve by following the flow of fluid in the pipe.
2. Consider each item of equipment affected by the event sequence and the latent hazards which can exist in each. These may be wrong state, wrong substances present, foreign materials, incorrect interconnections, valves being closed or open, etc. For mechanical systems, consider the effect of sticking, equipment being jammed, stick-slip possibilities, interferences, etc.

In many cases, the potential latent hazard can and should be checked by the operator. In this case, the error combination of interest will be the presence of the latent hazard together with the omission or misinterpretation of the check. In other cases the latent hazard will have been caused by an earlier error, such as failure to close the air hose valve in the example above. In some cases, the latent hazard will be unobservable, i.e. a perfect trap. In this case, no operator error is involved, only a design error.

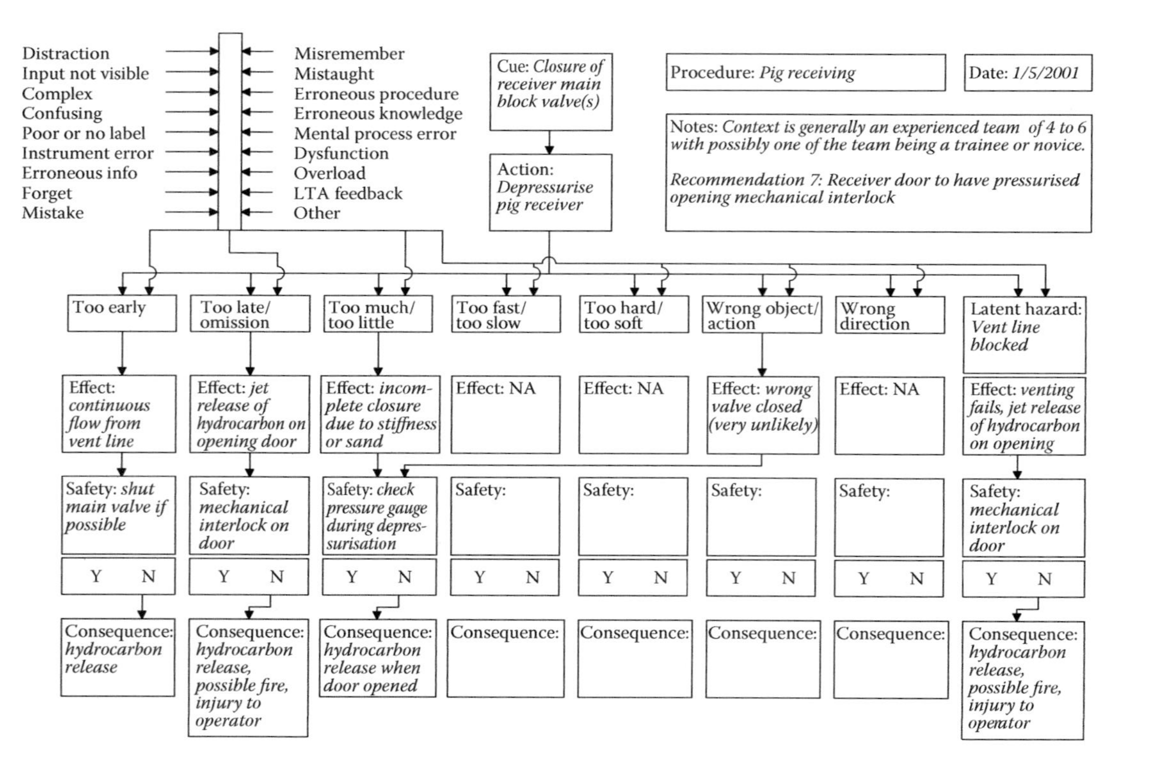

FIGURE 18.1 Template for AEA. One analysis sheet is needed for each task step. The sheet is filled out in Table 18.1. The cause–consequence diagram has some features which help to derive a reasonably accurate analysis. The step being analysed is written at the top of the cause–consequence diagram. Just above it is the 'cue', that is, the signal which sets a person on the path to performing a task or task step. The cue, for example, may be an alarm feedback which indicates completion of the previous step or may be self-cueing, that is, performing a step as soon as the previous step is completed. The cueing for a step is important, because poor cueing or distraction may make it likely that a step may be omitted.

ERROR CONSEQUENCES

For each error mode and each latent hazard identified, the possible consequences are evaluated. Standard techniques of cause consequence analysis can be used. If there is a potential for very serious consequences, the possibility of failures in the safety devices, as well as the initial error, should be taken into account.

If the error is serious, the analyst should ask the following:

1. Is the effect of the error observable?
2. Can the error be corrected?
3. Is there enough time to perform the correction?

These steps can be built into the cause—consequence diagram too. In the analysis template shown in Figure 18.1, provision is made for safety actions which may be operator response to the error (self-recovery) or may be automated safety features such as trips.

RECOVERY

If an error can cause a serious consequence, it is important to know whether the effects of the error are visible, or indicated by alarms, and whether there is both mechanism and time for the operator to recover. For each significant error the symptoms in terms of alarms, movement on trend charts, audible feedback, vibration and visible effects should be considered. The time available to correct the error or prevent serious effects should be estimated or calculated. Also, the possibility of automatic safeguards needs to be recorded.

ACTION ERROR CAUSE ANALYSIS

If an error has a serious consequence, it would be nice to know whether it is likely to occur. As will be seen in later chapters, this cannot be determined from the error mode alone. In fact, the probability of an error is (to a good approximation) the sum of the probabilities of the causes.

It would also be helpful to know how to prevent the error, or at least make it less probable. As will be seen, prevention of errors is largely a question of limiting or eliminating causes.

For each error mode, there is a set of typical error causes. For example, repairing the wrong pump (a frequent and often fatal error) is generally caused by poor labelling, poor specification of the maintenance work order or illogical order or position of the location of pumps. The pumps for tank A, for example, may be placed close to tank B, so that confusion is possible.

Over the years, many AEA different formats have been tried in order to be able to relate causes to errors in an effective way. With such a large set of possible causes, it is not easy to represent them all on one page. The most effective format that has been found for in-depth analysis is the generic cause consequence diagram shown in Figure 18.1, with a full range of error modes shown and with space for noting the most significant causes. A separate checklist of causes is provided in Table 18.1 so

TABLE 18.1
Error Cause Checklist

Error Mode	Cause Groups	Error Causes
Too early, premature action	Action prevented or hindered	Attention failure
Delay		Comfort call
Omission		Distraction
Too much force		Overload
Too little force		Priority error
Too fast		Action prevented, hindrance (locked doors or gates, areas closed off for safety reasons such as radiography or lifting operations taking place, known leaks)
Too slow		
Too much/too long		
Too little/too short		
Wrong direction	Observation error	Hidden cue, symptom
Wrong object		Mistaken cue, symptom
Wrong command		Cue overlooked
Wrong tool		Symptom overlooked
Wrong value		Missing cue
Wrong substance		Hidden symptom
Wrong location/path		Parallax
Wrong action		Mistake
Too early, premature action	Identification error	Reasoning error
Delay		Misremembering
Omission		Mislearning
Too much force		Erroneous or LTA training
Too little force		Ambiguous symptoms
Too fast		Missing symptoms
Too slow		Symptom confusion
Too much/too long		Misleading symptoms
Too little/too short		Misleading labelling
Wrong direction		Misleading text
Wrong object		Error in procedure
Wrong command		Communication error
Wrong tool		Hidden symptom
Wrong value		Information overload
Wrong substance		Change in system
Wrong location/path		Location error
Wrong action		Mistake
		Tunnel vision
		Fixation
	Decision error	Reasoning error
		Misremembering
		Mislearning
		Erroneous or LTA training
		Misleading information
		Missing information
		Communication error

(*Continued*)

TABLE 18.1 (CONTINUED)
Error Cause Checklist

Error Mode	Cause Groups	Error Causes
		Decision paralysis
		Diagnosis difficulty
		Erroneous model
		Incomplete model
		Inappropriate goal
		Decision paralysis
	Planning error	Reasoning error
		Misremembering
		Plan shortcut
		Experimentation
		Priority error
		Too short planning horizon
		Inappropriate goals
Too early, premature action	Execution error	Misremembering
Delay		Mislearning
Omission		Forget
Too much force		Lack of training
Too little force		Judgement error
Too fast		Mislearning
Too slow		Procedure confusion
Too much/too long		Error in procedure
Too little/too short		Shortcut
Wrong direction		Trapping error
Wrong object		Communication error
Wrong command		Timing error
Wrong tool	Motoric error	Stiffness
Wrong value		Jamming
Wrong substance		Tiredness
Wrong location/path		Overenthusiasm
Wrong action		Stick-slip effect
		Lack of training
		Change in system response
		Too high force required
		Too high speed required
		Too high precision required
		Reversion to stereotype
Omission	Distraction	Other tasks
		Telephone
		Colleague
		Boss
		Disturbance
		Equipment failure
		Personal business, worries

(*Continued*)

TABLE 18.1 (CONTINUED)
Error Cause Checklist

Error Mode	Cause Groups	Error Causes
Wrong action	Reasoning error	Side effects overlooked
Wrong procedure		Precondition overlooked
		Deduction error
		Error in mental model
		Error in mental simulation of effects or causes
		Overgeneralisation
		Erroneous analogy
Wrong action	Erroneous procedure	Omission of step
Wrong procedure		Incorrect action
		Omission of warning
		Error in text
		Mismatch with actual equipment
	Communication error	Ambiguous message
		Wrong terminology
		Terminology not understood
		Noise
		Incomplete message
		Erroneous message
		Mixed communications
	Decision paralysis	Stimulus overload
		Shock
		Fear
		Lack of knowledge
		Information overload
	Cue overlooked	Distraction
		Cue hidden
		Cue nonarrival
		Cue confusion
		Multiple cues
		Information overload
	Symptom overlooked	Distraction
		Symptom hidden
		Symptom nonarrival
		Symptom confusion
		Conflicting symptoms
		Instrument error
		Information overload
		Fixation
		Tunnel vision
	Hindrance	Equipment failure
		Countermand
		Localisation problem
		No light

(Continued)

TABLE 18.1 (CONTINUED)
Error Cause Checklist

Error Mode	Cause Groups	Error Causes
		Smoke
		Access difficulty
		Lack of air
		High temperature
		Hazardous material
		Radiation
		Flooding
	Access difficulty	Fire
		Distance
		Height
		Obstructing equipment
		Narrow access
		Locked door
		Jam

that it is not necessary to include all of the causes on one page. This approach is useful also, because many of the causes, such as lack of standard operating procedures and complexity, apply to a whole task and not just a particular step.

RISK ANALYSIS OF THE ERRORS

The risk from any error is the total frequency with which the procedure or activity is undertaken multiplied by the probability of human error multiplied by the probability of failure to recover multiplied by the probability of failure of automated safety devices:

$$\text{frequency of accident} = \text{frequency of activity} \times \text{human error probability} \times (1 - \text{probability of recovery}) \times \text{probability of failure on demand of safety devices.}$$

Methods for obtaining the probability of error are described in Chapter 19.

Risk is a function of accident frequency and consequence. Risk analysis can be made using the methods described in Chapter 20.

RISK REDUCTION

The purpose of the AEA is to reduce risk. Accident prevention must generally be based on the causal analysis—remove the cause and you reduce the risk. Methods of risk reduction are described in Chapter 21.

There will generally be many methods which can be used for risk reduction. Making preventive measures for each error cause, or even for each error mode, can lead to a patchwork of measures with no clear philosophy. When the analysis is completed, it is preferable to stand back from the analysis and choose an overall approach which is consistent and achieves the needed risk reduction in a coordinated way.

EXAMPLE OF ACTION ERROR ANALYSIS—REMOVING A SCRAPER FROM A SCRAPER TRAP (PIG RECEIVER)

Pipelines need periodic cleaning to remove water and corrosion scale, and in the case of gas pipelines, to remove liquids. If scraping, or 'pigging' as it is also known, is not done, sediment or water can cause increased corrosion and blockage in extreme cases. The scraper or 'pig' is a plug or ball which is forced through the pipe by the pressure of liquid or gas behind it (Figure 18.2). It may take up to several days for a pig to pass from a launcher to a receiver, although for short pipelines a period of just a few hours is more typical.

Pig receivers are long vessels, placed at the end of a pipeline (or sometimes along a pipeline). It is usually isolated from the pipeline so that it is not pressurised, but when a pig is being sent, valves are opened to allow oil or gas to enter. Figure 18.2 shows a typical pig receiver, from an oil pipeline.

The pig is pushed through the pipeline by the pressure of oil (or gas) pumped into the pipeline. The oil ahead of the pig passes through the receiver bypass valve to whatever vessel is designated to receive the oil. This may be a slug catcher, that is, a large vessel with sufficient capacity to take the mixed water and oil from the pipeline, which can then be transferred more slowly to oil processing, so that water can be separated from the oil.

During the first part of the pigging process, when relatively clean oil is being pushed ahead of the pig, the pig receiver's bypass valve remains open and the pig receiver's block valves are closed. When the pig approaches the receiver, the valve into the receiver and the valve from the receiver (kicker line valve) are opened, and the bypass valve closed. When the pig passes into the receiver, the bypass valve can again be opened, and the receiver valves closed. The receiver can then be depressurised and drained, the door opened and the pig removed, along with any sludge, which is collected in a container.

The procedure is hazardous, firstly, because of the pressures and, secondly, because highly flammable crude oil, for example, may be freed to the atmosphere when the pig receiver is opened.

Latent hazards identified for the scraper receiver were the following (Tables 18.2 through 18.9):

1. Leak present at isolation valve opening
2. Blockage in drain line
3. Broken internals on block or bypass valves
4. Failure of vent line valves to open
5. Too late start of the procedure
6. Failure to ensure that the vent and drain valve are closed
7. Too rapid opening of the receiver isolation valve
8. Failure to open either of the receiver valves
9. Too late start of the procedure

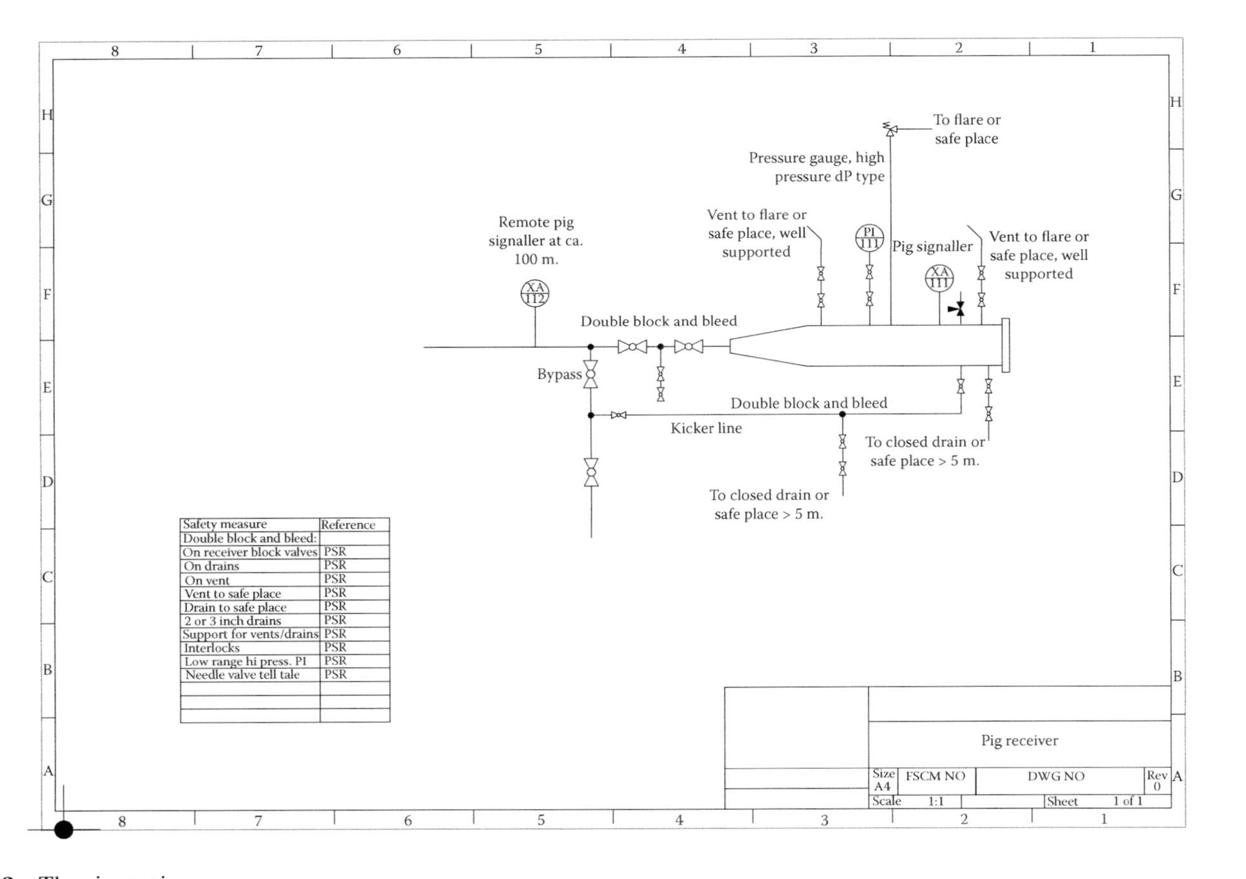

FIGURE 18.2 The pig receiver.

TABLE 18.2
Pig Receiver AEA: Sheet 1

Action Error and Latent Hazard Analysis

System: Pigging system | **Subsystem:** Pipeline | **Author:** JRT | **Date:** 28/09/2003
Item/Node: Pig launcher | **Procedure:** Pig receiving | **Step:** Initial system state

Disturbance, Causes	Consequences	Safety Measures	Recommendations	F	B	C	R
1. Leak/pig launcher • Corrosion • Material failure • Door not fastened tightly, seal leak • Drain valve left open	• Release of oil: 1 • Pollution • Jet fire: 2 • Domino effects: 3	1. ESD 2. Low ignition probability 3. Firefighting systems		–2	3	8	5
2. Rupture/pig launcher • Corrosion • Material failure • Locking failure • Opening without depressuring: 1, 2 • Opening with inlet valve leaking or open • Opening with back pressure from system • Leaks from the rest of the system • Back pressure from downstream: 3 • Back pressure from closed drain system: 4, 5 • Cross flow due to valve leak from other equipment	• Release of oil: 6 • Pollution • Fire: 7 • Jet fire • Domino effects: 8 • Flash fire: 9	1. Interlock 2. Pressure locking door 3. Check valve 4. Block valve closed according to procedure 5. Check valve 6. ESD 7. Low ignition probability 8. Firefighting systems 9. ESD		–3	4	8	1

Note: For Tables 18.2 through 18.9, the headings of the fifth to eighth columns are as follows: F—frequency, given as log of occurrence per year, e.g. –3 is 1 per thousand years; B—barrier risk reduction, given as the log of the risk-reduction factor, e.g. 3 is a factor 100; C—consequence, given as log of the 'damage', e.g. 7 corresponds to $10,000,000 and R—index of risk, or logarithm of the loss per year in dollars. Numbers in these columns indicate safety measures. Numbers in parentheses correspond to residual risk after implementation of recommendations. ESD: emergency shutdown.

TABLE 18.3
Pig Receiver AEA: Sheet 2

Action Error and Latent Hazard Analysis

System: Pigging system | **Subsystem:** Pipeline | **Author:** JRT | **Date:** 28/09/2003
Item/Node: Pig receiver | **Procedure:** Pig receiving | **Step:** Initial state

Disturbance, Causes	Consequences	Safety Measures	Recommendations	F	B	C	R
3. Leak/pig receiver • Corrosion • Material failure • Door not fastened tightly, seal leak • Drain valve left open	• Release of oil: 1 • Pollution • Jet fire: 2 • Domino effects: 3	1. ESD 2. Low ignition probability 3. Firefighting systems		–2	3	8	5
4. Rupture/pig receiver • Corrosion • Material failure • Locking failure • Opening without depressuring: 1, 2 • Opening with inlet valve leaking or open • Opening with back pressure from system • Leaks from the rest of the system • Back pressure from downstream: 3 • Back pressure from closed drain system: 4, 5 • Cross flow due to valve leak from other equipment	• Release of oil: 6 • Pollution • Fire: 7 • Jet fire • Domino effects: 8 • Flash fire • Operator injury: 9	1. Interlock 2. Pressure locking door 3. Check valve 4. Block valve closed according to procedure 5. Check valve 6. ESD 7. Low ignition probability 8. Firefighting systems 9. ESD		–3	4	8	1

Note: See note to Table 18.2 for the key to the headings of and entries in the fifth to eighth columns.

TABLE 18.4
Pig Receiver AEA: Sheet 3

Action Error and Latent Hazard Analysis

System: Pigging system | **Subsystem:** Pipeline | **Author:** JRT | **Date:** 28/09/2003
Item/Node: Pig receiver | **Procedure:** | **Step:** 1. Prepare for pig receiving

Disturbance, Causes	Consequences	Safety Measures	Recommendations	F	B	C	R
5. Action with latent hazard/ prepare for pig receiving • Latent hazard in pipe • Leak in drain line • Leak at door	• Possible release when isolation valve is opened • Fire: 1 • Injury to operator • Spill	1. Low ignition probability	1. Open isolation valve slowly and check for pressure rise. 2. Stand clear of potential leak points when opening isolation valve.	−2	2	8	4
6. Action with latent hazard/ prepare for pig receiving • Latent hazard in pipe • Blockage in drain line	• No draining • Release when door is opened: 1 • Drain may remain opened while receiver pressurised • Sudden unblocking when trap is pressurised • Fire: 2 • Injury to operator • Spill	1. Visual check of drain outflow 2. Low ignition probability	3. Open isolation valve slowly and check for pressure rise. 4. Stand clear of potential leak points when opening isolation valve. *Note:* Visual check of drain outflow is possible only if the drain is an open one.	−3	2 (4)	8	3 (1)
7. Omission/ensure that drain valve and vent valve are closed • Operator error • Confusion of responsibilities • Excessive haste	• No draining • Release when door is opened: 1 • Drain may remain opened while receiver pressurised • Sudden unblocking when trap is pressurised • Fire: 2 • Injury to operator • Spill	1. Visual check of drain outflow 2. Low ignition probability		−3	2	8	3

Note: See note to Table 18.2 for the key to the headings of and entries in the fifth to eighth columns.

TABLE 18.5
Pig Receiver AEA: Sheet 4

Action Error and Latent Hazard Analysis

System: Pigging system | **Subsystem:** Pipeline | **Author:** JRT | **Date:** 29/09/2003

Item/Node: Pig receiver | **Procedure:** | **Step:** Open isolation valve bypass

Disturbance, Causes	Consequences	Safety Measures	Recommendations	F	B	C	R
8. Omission/open isolation valve bypass • Operator error • Unfamiliarity with procedure • Training on pig receiver without bypass	• Possible hammer on opening isolation valve • Possible rupture		5. Fit pressurisation interlock. *Note:* Hammer is possible only for smaller traps—for large traps isolation valve opening is slow.	–3	(3)	8	(2)
9. Too fast/open the receiver isolation valve in the circulation line around the receiver isolation valve to pressurise the receiver • Excessive haste	• Hammer • Potential rupture		*Note:* This is possible only for smaller traps.	–4		8	4
10. Omission/open the receiver isolation valve in the circulation line around the receiver isolation valve to pressurise the receiver • Operator error • Confusion in procedure	• Pig will not enter trap • Problem of pig jamming in valves • Possible later catastrophic release		6. Check for pig entry to receiver. 7. Valve full opening interlock.	–3	(2)	8	(3)

Note: See note to Table 18.2 for the key to the headings of and entries in the fifth to eighth columns.

TABLE 18.6
Pig Receiver AEA: Sheet 5

Action Error and Latent Hazard Analysis

System: Pigging system | **Subsystem:** Pipeline | **Author:** JRT | **Date:** 29/09/2003
Item/Node: Pig receiver | **Procedure:** | **Step:** Open the receiver isolation bypass valve

Disturbance, Causes	Consequences	Safety Measures	Recommendations	F	B	C	R
11. Too little/open the receiver isolation valve • Operator error • Mistake • Excessive haste • Valve sticks	• Pig sticks at a later stage • Catastrophic pig release when door is opened		8. Valve full opening interlock.	−3	(3)	8	(2)
12. Omission, too late/close the bypass valve • Operator error • Forget	• Possible release later when door opened: 1	1. Depressurisation checks		−3	2	8	3
13. Omission/open the receiver discharge valve • Operator error	• Pig may not enter trap • Possible hammer when main line valve closed			−4		8	4
14. Omission/close the main line valve • Operator error	• Pig may be sucked into main line flow • Possible pig entry to pump		9. Fit an interlock for full main line valve closure.	−2	(3)	8	6 (3)
15. Too little/close the main line valve • Operator error • Valve sticks	• Pig may be sucked into main line flow • Possible pig entry to pump		10. Fit an interlock for full main line valve closure.	−3	(−3)	4	1 (−2)

Note: See note to Table 18.2 for the key to the headings of and entries in the fifth to eighth columns.

TABLE 18.7
Pig Receiver AEA: Sheet 6

Action Error and Latent Hazard Analysis

System: Pigging system | **Subsystem:** Pipeline | **Author:** JRT | **Date:** 29/09/2003
Item/Node: Pig receiver | **Procedure:** | **Step:** Check that pig signallers are down

Disturbance, Causes	Consequences	Safety Measures	Recommendations	F	B	C	R
16. Omission/check pig signallers are down • Operator error	• No indication of pig position • Pig may remain in line • Blockage problems during recovery • Possible catastrophic release • Pig may become stuck in isolation valve • Possible catastrophic release when door opened		11. Use reliable noncontact pig signallers. 12. Provide full closure indicator for isolation valve.	–3	(4)	8	5 (1)
17. Premature/wait until the pig arrives • Signaller not operational	• Pig may remain in line • Blockage problems during recovery • Possible catastrophic release		13. Use reliable noncontact pig signallers. 14. Provide full closure indicator for isolation valve.	–3	(4)	8	5 (1)
18. Omission/open the main line valve • Operator error	• Loss of flow • No progress, corrected when lack of progress is noticed						0
19. Omission/close the receiver isolation valve • Forget • Serious confusion of procedure	• Pig trap remains pressurised: 1 • Release when door opened	1. Depressurisation checks	15. Fit an isolation valve closure interlock. 16. Fit depressurisation interlock for door.	–3	(4)	8	5 (1)

Note: See note to Table 18.2 for the key to the headings of and entries in the fifth to eighth columns.

TABLE 18.8
Pig Receiver AEA: Sheet 7

Action Error and Latent Hazard Analysis

System: Pigging system | **Subsystem**: Pipeline | **Author:** JRT | **Date:** 29/09/2003
Item/Node: Pig receiver | **Procedure:** | **Step:** Close the receiver isolation valve

Disturbance, Causes	Consequences	Safety Measures	Recommendations	F	B	C	R
20. Too little/close the receiver isolation valve • Valve sticks	• Pig trap remains pressurised: 1 • Release when door opened	1. Depressurisation checks		–3	(4)	8	5 (1)
21. Omission/close the receiver discharge valve • Forget • Serious confusion of procedure	• Pig trap remains pressurised: 1 • Release when door opened	1. Depressurisation checks	17. Fit a discharge valve closure interlock. 18. Fit depressurization interlock for door.	–3	(4)	8	5 (1)
22. Too little/close the receiver discharge valve • Valve sticks	• Pig trap remains pressurised: 1 • Release when door opened	1. Depressurisation checks	19. Fit a discharge valve closure interlock.	–4	(4)	8	4 (0)
23. Action with latent hazard/open the vent valve • Drain line blocked • Pressure sensor failed	• Receiver remains pressurised: 1 • Accident when the door is opened		20. Provide spare pressure sensor and repair procedure for the case where the pressure gauge fails to register.	–3			0
24. Forget/open the drain valve • Operator error	• Receiver remains filled: 1 • Accident when the door is opened, possible fire		21. Take care when opening the door; crack it open before opening fully.	–3	(–1)	7	4 (3)

Note: See note to Table 18.2 for the key to the headings of and entries in the fifth to eighth columns.

TABLE 18.9

Pig Receiver AEA: Sheet 8

Action Error and Latent Hazard Analysis

System: Pigging system | **Subsystem:** Pipeline | **Author:** JRT | **Date:** 29/09/2003

Item/Node: Pig receiver | **Procedure:** Pig receiving | **Step:** Open the vent valve

Disturbance, Causes	Consequences	Safety Measures	Recommendations	F	B	C	R
25. Omission/open the vent valve • Operator error	• Reduced checking		*Note:* The check should be by listening for depressurisation flow. There is a potential conflict between depressurisation via the vent and draining.				0
26. Action with latent hazard/open the receiver door • Receiver still pressurised • Pig stuck in isolation valve	• Major release when door opens • Injury to operator opening door		22. Fit a door opening interlock, based on depressurisation	–3	(3)	7	4 (1)
27. Action with latent hazard/extract pig • Sulphide scale present	• Fire due to autoignition: 1	1. Wetting the scale immediately	Ensure that there is water available (hose connection and bucket).	–1	(1)	4	2
28. Mishandling/extract pig	• Pig drops • Ignition due to friction • Fire			–3		2	–1
29. Not enough/close and lock the receiver door • Mechanical damage • Dirt in the mechanism	• Possible door rupture on pressurisation • Injury: 1	1. Persons unlikely to be in line	23. Fit a door closure interlock.	–4	–1	8	3

Note: See note to Table 18.2 for the key to the headings of and entries in the fifth to eighth columns.

10. Failure to close the bypass valve once the pig is in the receiver
11. Failure to close the receiver block valves once the pig is in the receiver
12. Failure to depressurise or drain the pig receiver before opening the pig receiver door
13. Failure to close the drain or vent valve after opening the door
14. Presence of pyrophoric iron sulphide

Errors of this kind can cause dramatic effects and accidents. Just one of these will be discussed here, as an illustration.

Error 12, too late or omission of closure of the bypass, will lead to the pig being forced sideways into the bypass valve. This error has occurred many times and usually results in the pig sticking in the line. The pipeline must then be shut down and drained, flushed with nitrogen to prevent ignition and explosion; the valve dismantled; the pig removed; the line flushed with nitrogen (again) and the system reassembled and restarted. In a large crude oil line, this can typically take 2 days or more and can cost up to $100 million. Fortunately, in most cases the cost is less than this. The tee junction ahead of the scraper is in most installations implemented as a 'barred tee'; that is, it has a blocking bar welded across the opening to prevent the scraper being forced into the bypass line. This measure is not always successful, however. The author has investigated several cases in which the scraper was forced past the tee due to the timing error plus a high pressure drop.

The frequency of this error occurring must be determined from the causes. Several causes are possible:

1. The pumping occurs at a higher rate than expected, so that the pig arrives early.
2. The calculation of pig arrival time is in error.
3. The pig arrival indicator has failed (contact types are notoriously unreliable; in modern designs noncontact pig detectors are used) (latent hazard).
4. The operators are delayed or prevented from arriving (e.g. truck failure) and compound the problems by not informing the control room so that the pumps can be stopped.
5. The operators simply misunderstand or forget the necessity for timely operation.

It can be seen that these causes are of completely different kinds. The frequency of the failure mode can be determined only at the detailed level, for each individually.

RISK REDUCTION FOR THE SCRAPER TRAP OPERATION

The most important point in any analysis is the impact it will make on risk reduction. The analysis above shows the importance of several design features:

1. The sequencing of valve opening and closure
2. The reliability of the pressure gauge
3. The tendency of the drain line to block
4. The tendency for the isolation valves to leak

5. The reliability of the pig arrival sensors
6. The position of persons when the door is opened
7. The location and direction of the depressurisation vent
8. The possibility of any back leakage via the drain line
9. The degree of knowledge and hazard awareness of the operators and, particularly, of the operating team supervisor

All of these things can be guarded against. For point 1, for example, mechanical interlocks can be provided which force a particular opening and closure sequence. It is also usual in modern designs to select a receiver door type which cannot be opened under pressure because there is too much resistance to unlock the door. These approaches have become standard practice in many companies, in some of them as a direct result of the AEA results.

For several of the errors, the procedure can be improved by adding extra stages of checks and explanations of the reasons for these.

TRACKING OF OPERATIONAL STATE AGAINST A MENTAL MODEL

It is possible to carry out the pig receiver task just by following the instructions in the operating procedure. An experienced operator, though, is watching for the arrival of the pig. If the pig detector fails, he or she may in any case detect the arrival by the sound and because of the expectation of the timing. At the depressurisation the operator watches for the fall on the pressure gauge, confirming depressurisation. He or she watches particularly to see the movement of the indicator.

In other words, the experienced operator has a model of the process in mind (or one of his or her minds, because the model is used for mental simulation of the process, mostly subconsciously). The experienced operator keeps the mental simulation synchronised with what is happening in the plant, by observing the small clues from the instruments, changes in sound of flows and changes in appearance of pipe surfaces. These are clues which allow the operator to 'see through steel'.

The difference between the two cases of simply following procedures and of tracking the process makes a large difference to the probability of error. The process-tracking mode of operation especially allows the detection of latent hazards if these are in any way 'visible'. The model-tracking mode does not seem to make a major difference to the reliability of actually *performing* procedural steps.

There are large differences between the extents of process tracking which operators and designers undertake. The author recorded operator descriptions of process behaviour during HAZOP studies and found very large differences in degree of knowledge and hazard awareness. The probability of error will correspondingly vary considerably.

OBSERVATIONS ON THE EXAMPLE ANALYSIS

The detailed analysis of error causes in the previous section covers just one error mode, error 12, on page 222 but is nevertheless quite extensive. An analysis at this depth is desirable, since any of the aspects described above can change the probability of an accident by two orders of magnitude.

The analysis of error causes is not too obvious in the preceding pages. The reason for this is that there are enough case histories of incidents to allow frequencies to be determined without a detailed causal analysis.

It is worth noticing the extent of the analysis. In a complete risk analysis of a full process plant, an analysis at this depth, with full quantification of frequencies, is not possible in oil and chemical industry. Such analyses are carried out for nuclear power plants and other systems with long design lead times. It would imply several thousand pages of text. The method can definitely be applied though to typical designs, and the lessons learned from these can be applied uniformly in an installation. The practical approach to this problem is to provide a library of generic analyses, as has been done in Ref. [12].

REFERENCES

1. S. J. Munger, R. W. Smith, and D. Payne, *An Index of Electronic Equipment Operability: Data Store AIR-C43-1/62-RP(1)*, Pittsburgh, PA: American Institute for Research, 1962.
2. A. D. Swain and H. E. Guttman, *Handbook of Human Reliability Analysis with Emphasis on Nuclear Power Plant Applications*, Sandia National Laboratories, 1983, published as NUREG/CR-1278, U.S. Nuclear Regulatory Commission.
3. B. Kirwan, *A Guide to Practical Human Reliability Assessment*, London: Taylor & Francis, 1994.
4. J. C. Williams, HEART – A Proposed Method for Assessing and Reducing Error, 9th Advances in Reliability Symposium, University of Bradford, 1986.
5. J. R. Taylor, *Interlock Design Using Fault Tree and Cause Consequence Analysis*, Risø National Laboratory Report M-1890, 1976.
6. J. R. Taylor, *Risk Analysis for Process Plant, Pipelines and Transport*, London: E & FN Spon, 1994.
7. J. R. Taylor, O. Hansen, C. Jensen, O. F. Jacobsen, M. Justesen, and S. Kjærgaard, *Risk Analysis of a Distillation Unit*, Risø National Laboratory Report M-2319, 1982.
8. B. Kirwan and L. K. Ainsworth, *A Guide to Task Analysis*, Boca Raton, FL: CRC Press, 1992.
9. J. Annett and K. D. Duncan, Task Analysis and Training Design, *Occupational Psychology*, Vol. 41, pp. 211–221, 1967.
10. N. A. Stanton, Hierarchical Task Analysis: Development, Applications and Extensions, *Applied Ergonomics*, Vol. 37, pp. 55–79, 2006.
11. D. Embrey, *Task Analysis Techniques*, Wigan, UK: Human Reliability Associates, 2000.
12. J. R. Taylor, A Catalogue of Human Error Analyses for Oil, Gas and Chemical Plant, http://www.itsa.dk, 2015.

19 Error Probabilities

WHY DO WE NEED TO QUANTIFY ERROR PROBABILITY?

Given that human errors in operation can be predicted to a large extent, why do we not just eliminate the error causes? This is in fact a good opening strategy which can be very successful, as can be determined by comparing the frequencies of accidents in the best and the worst companies. In the end though, there is a limit to the resources, in particular human resources, which can be dedicated to error reduction.

As an example, should a company introduce simulator training for all its units, some of them or none of them? Given that such a simulator has a typical cost of about $2 million, will require at least 2 weeks training for each operator and will require hard work for a trainer and a plant specialist for 6 months to a year, should the time and effort be used here or in some better way? Or, to take another example, should automatic safety systems be implemented to prevent release of oil to the sea for a system which has never done so in over 40 years of operation, given that a good oil spill emergency plan is already implemented, with a good provision of booms, recovery pumps, etc.?

The answer to the question of how many human error measures to apply is today answered in one of three ways. The first is by standard—there are several operator safety guidelines and standards available today. The second approach is to make a risk assessment and then reduce the resulting risk to the as low as reasonably practicable (ALARP) level. This approach is taken in the nuclear power industry. (Risk analysis for safety engineering is also widely applied in the oil and chemical industries, but human error risk analysis is only rarely made.)

The third approach is to wait until an accident occurs and then reduce the risk.

CASE HISTORY 19.1 A Pragmatic Solution Based on Experience

A chemical batch reactor producing pharmaceutical ingredients lost a complete batch due to the bottom drain valve being left open. The loss was worth about $20 million. The company then installed two position switches (one to signal closed valve and the other to signal a fully open valve) on all its operational drain valves. Placing limit switch pumping interlocks on manual valves is an unusual design practice and would be expensive if applied generally. It is doubtful whether a company could be persuaded to implement such a program without the object lesson of a major dollar loss, no matter how well the risk analysis demonstrated the need. But it is worth trying.

RISK DETERMINATION BASED ON STATISTICS

For some of the error causes described in the earlier chapters, detailed quantification of impact is very difficult. For example, what is the impact of providing enhanced operator training? Quantification of this at the detailed level may be feasible for a few accident scenarios using the techniques shown in later sections of this chapter. Quantifying the effect in detail for a complete refinery would hardly be feasible, at least with the techniques available today. (Extensive human reliability analysis is carried out for nuclear power plants, but the timescale for construction of such plants is many years, and the value of the plants is very high.)

Large accident reports and statistics can be used as a basis for quantifying impact on risk, at least for some accident types. Figure 19.1 gives a summary of accident causal factors taken from the U.S. Chemical Safety Board (CSB) and United Kingdom's Health and Safety Executive accident reports. In all, 130 large accidents were considered. It has not been possible to identify the causes of all of these, but for the 45 where assessments of cause have been published, the distribution of possible methods of risk reduction was as in Figure 19.1.

As can be seen, working on operator errors, maintenance errors and procedures has a chance of preventing over half of the accidents.

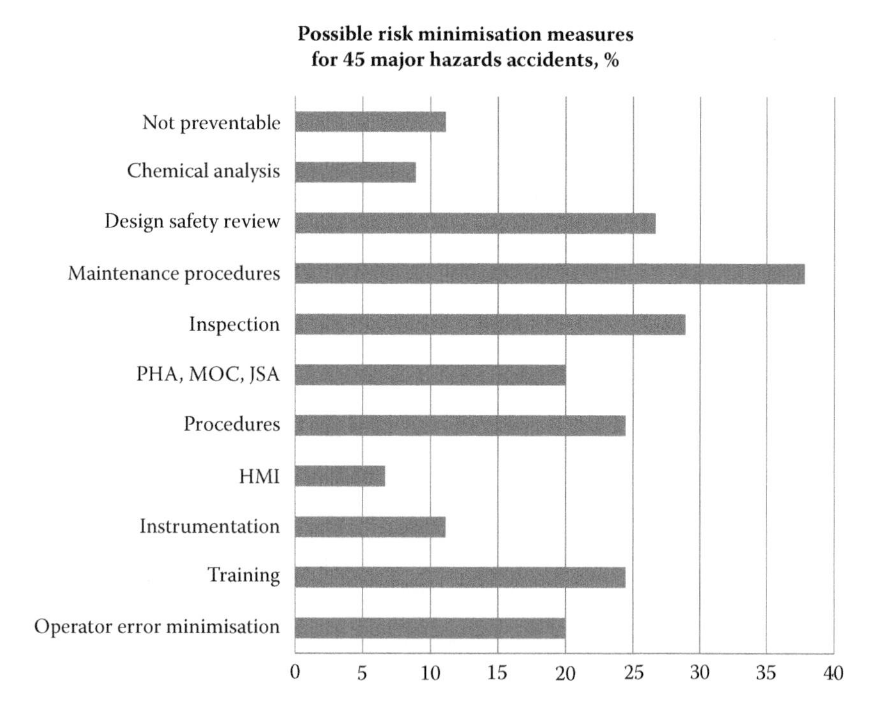

FIGURE 19.1 Possible ways in which risk could be reduced for major accidents. Note that percentages add to more than 100 because for many accidents, more than one technique could be used. HMI: human–machine interface; JSA: job safety analysis; MOC: management of change; PHA: process hazard analysis.

The actual levels of risk at the installation level are difficult to determine for most installation types and most countries. Estimation is possible for U.S. refineries because, firstly, the actual installations are well documented and the accident reports well published.

According to Marsh [1], property loss for refineries around the world from 2000 to 2013 was about $15 billion. Of this, about half was in the United States. This gives about $1 billion per year of property losses due to large accidents alone. Our calculations of business interruption losses and liability compensation gives a value two to three times larger (not taking into account exceptional cases with very large liability compensation). With about 140 refineries in United States, this gives a loss from major accidents of about $2 billion to $3 billion per year on average. Minor and medium-size accidents add significantly to this level of risk, since, although the loss from each accident is less, the accidents occur more frequently. It is, therefore, worthwhile investing at least this amount in risk reduction efforts each year, beyond that which is currently made.

Actually, this calculation underestimates the justifiable investment for some refineries, since most of the accidents occur with the poorer performers. The better performers are presumably already putting more effort into risk reduction.

This calculation is a purely economic one, concerning the potential losses and the protection of shareholder value. In addition, there are several lives lost each year, sometimes tens of lives, and many large accidents have a considerable impact on the public.

DETAILED ERROR PROBABILITY DETERMINATION

In order to determine in detail the level of risk arising from human error, it is necessary to determine which errors can occur, the frequencies to be expected for the errors, the reliability of any safety measures protecting against error or error consequences and the extent of those consequences. Predicting the frequency or probability of error is an important step.

In principle, human reliability data collection can be carried out in a similar way to equipment reliability analysis:

1. Observe how many opportunities arise.
2. Observe how many errors occur.
3. Determine the mode and cause of the error.
4. Divide the number of errors by the number of possibilities of error.

In practice, it is not so easy to do this. Firstly, it is not so easy to observe errors. Human errors disappear as soon as they are committed, and only the consequences remain. Even the consequences disappear in most cases, because the majority of errors are actually corrected. Only if the results are unrecoverable are the data available. As a result, in the field only about 1 in 100 errors are observable.

Secondly, such observation takes a long time. With typical error rates of 1 per 1000 to 1 per 10,000 actions, it generally takes at least a week to observe just one error. To observe the more seldom error types can take a year or more. The observation needs to be very close, because most errors are corrected very rapidly. In addition, very close observation is disturbing for many operators, at least at first. In any case, one must surmise that observation changes behaviour. In fact, if you know very much about the plant, it is almost impossible to observe closely without both transferring knowledge to the operators and serving as a check to prevent the errors with the work consequences.

Thirdly, the causes, and to some extent the modes of error, are not directly observable. In many cases the cause can be determined only by in-depth root cause analysis or interviews after the event. Such interviews need to be made in an atmosphere of no-blame culture, and with sympathy, and usually need to take place over a cup of coffee or similar in order to get good information.

Fourthly, many of the error types of most interest are very rare, so that an enormous amount of observational data is needed before significant results can be obtained. (Rare error types are interesting, and data need to be collected for these, because there are so many different 'rare' types that overall, they are relatively frequent.)

Because of these problems, observation is difficult except in certain very repetitive situations, such as production line work. Direct observation in process plants is extremely difficult in a meaningful way, not least because operator behaviour changes when the operators are being watched.

Despite these difficulties, it has been possible over the years to obtain several thousand hours of observation in the control room to support the present modelling. This is sufficient at least to determine the typical frequencies of different activities in the control room.

A number of more advanced experimental techniques have been developed to supplement direct observation for error probability determination:

1. Operator logs and plant trend records can be reviewed together with the operator. This is a good discipline for an operation supervisor in any case, because the logs and records tell a great deal about difficulties arising in operations and can be used to improve plant performance.
2. Process simulators, developed for operator training, can be instrumented to such an extent that operator performance can be recorded. Analysis of the results is arduous. There are nevertheless problems with simulator recording, most importantly that it is difficult to get experienced operators to take part in such studies for long periods. It is easier to get information from training simulators, but the error rates during training are not really representative of error rates during normal operation. Tracking the development of operating accuracy during training can definitely be used to derive normal operations' error rate. These have been observed to have good predictive value.
3. Accidents, incidents and near misses are recorded carefully in most modern well-managed plants. Although operator errors may sometimes be suppressed in incident reports, most reports reviewed by the author seem fairly complete. Determining error probabilities from incident reports requires that the number of opportunities for effort and also the number which are

recovered be determined. This can be difficult unless the analyst is in fact part of the process plant operation team or carries out extensive on-site analysis. There is a methodological problem in studying human error rates on the basis of accident data, that is, that accidents occur only for those errors which are not recovered, either by correction at the operator or by automatic safety systems.

4. There is one way in which accident data can be used systematically to check error rates, that is, first to use available generic data in probabilistic risk assessments. Then the resulting calculations are checked against actual accident records for similar kinds of plants. This has been done for several accident types.
5. A method called 'accident anatomy' [2] was developed in order to extract the maximum possible information from accident reports. This method involves selecting a group of accident types, such as opening a pressurised vessel, making a cause–consequence diagram for the error causes and error consequences and then marking all accident reports by highlighting lines on the diagram for each accident. In this way the relative frequencies of causes and consequences can be determined by counting how many highlights there are on each branch.
6. It is possible to observe some of the more common error types by looking explicitly for them. This approach takes enormous patience, waiting for errors, and requires that many procedure executions are observed. It also requires that the operation team become accustomed to the analyst.
7. These difficulties are presumably the reason why there are very little actual observed human error data published.
8. Several other approaches to applying judgement to human error analysis have been described (Ref. [9]). Most of these require reliance on 'expert judgement' and availability of a number of experts with sufficient expertise to make such judgements. Such procedures are quite valid provided that the expert is properly calibrated by including in the judgement process some items for which the error rate is known from statistical data collection. Validation studies for such approaches indicate an obvious truth. The result of judgements depends heavily on the experience of the experts, and it is extremely difficult to get even consistent results, let alone ones which agree with observation.

There is one way in which accident data can be used systematically to derive error rates, that is, first to use available generic data in probabilistic risk assessments. Then the resulting calculations are checked against actual experience. This means that the best way of evaluating data is to collect it in association with making a risk analysis.

One use of judgement is to allow transfer of information about performance-influencing factors. For example, factors concerning low light levels can reasonably be transferred from studies of one type of assembly task to another provided that the analogy is sufficiently close. Judgement here concerns the closeness of analogy. This kind of judgement can be quite objective and should be documented explicitly.

Another type of judgement can be applied systematically to determination of error cause probabilities. Relative judgements on the frequency of errors or error causes can be made on the basis of general experience. To make such judgements in a systematic way, at least two error types with known probabilities are needed. One should have a high probability and one a low probability. These are called anchor points. The probability of other errors can then be judged relative to the anchor points. Judgement can be applied quite objectively if a sufficiently large collection of accident and incident reports is available. While such a collection rarely gives a good basis for absolute error frequency determination, it does give a basis for relative frequency determination (see Chapter 1).

Several other approaches to applying judgement to human error analysis have been described (see the Bibliography). Most of these require reliance on expert judgement and availability of a number of experts with sufficient expertise to make such judgements. Such procedures are quite valid provided that the expert is properly calibrated by including in the judgement process some items for which the error rate is known from statistical data collection. For practical purposes, the procedures given above are usually sufficient and have been applied in the data collection given in this chapter.

FREQUENCY OF LATENT HAZARD EVENTS

Latent hazard events require a combination of an operation and the presence of a latent hazard. The frequency of such events is then calculated as the frequency of the normal action being performed multiplied by the probability of presence of the latent hazard. For example, the frequency of release of hydrocarbon from a filter can be calculated as the frequency of filter change (e.g. once every 3 months) multiplied by the probability that there is a leak on one of the filter isolation valves (typically between 0.1 and 0.01). This gives an accident frequency for this case of 0.4 to 0.04 per year. For this reason, filters are often fitted with pressure indicators or double block and bleed valves (two valves with a vent or drain in between), so that if a valve leak does occur, the liquid pressure will not reach the filter.

ANCHOR POINTS

It is much easier to determine the distribution of error modes and causes from process incident records than it is to determine probabilities. This is because error causes are often registered in openly published accident and incident registers, so that the number of occurrences and sometimes the causes can be found, whereas the number of opportunities for error and the number of earlier recoveries are generally not mentioned and presumably not even determined.

A way around this is to find some kinds of errors for which the frequencies *can* be determined. These are called anchor points. The probabilities for other error types can then be determined from the ratio of the occurrences and from the anchor point probabilities, along with some general knowledge about the frequency of various operations.

This approach is not without problems. The number of opportunities for error and the potential for recovery differ according to failure cause. As a result, interpretation is needed. Nevertheless, the approach gives access to a large volume of realistic and relevant failure probability information.

Several anchor points have been given by Kirwan [3]. Some new anchor points are given in the following sections.

ANCHOR POINT DATA—EXAMPLES

It has been possible to check values of some activities in the control room and in the field, with sufficient information to provide a number of good anchor points.

An Error of Omission—Too Late Check

Alkyd resin, used for making paint, is made by reaction on oil, such as fish oil with phthalic acid, by heating the two substances together with a solvent in a kettle reactor. The reaction takes about 15 hours, and at the end the reaction proceeds enough for the viscosity to increase. There is no easy way to monitor the viscosity automatically, and in practice the end of the process is judged by eye. The resin is flowed down to a vessel containing cool solvent, and the reaction stops.

The work for this resin production was controlled by two senior technicians, working shifts. Apart from holidays (during plant shutdown) the two controlled eight batches per day, for 22 years. In that time, checking of batches was forgotten or delayed twice. Delaying cooling of a batch is a memorable event—the resin sets solid and must be removed from the reactor with a pneumatic drill. The work takes about 2 days, and the loss is several tens of thousands of dollars.

In all, over a period of 22 years, two batches were lost due to delay in checking the viscosity. There is about a 20-minute margin for the checking. The technicians must be regarded as showing considerable expertise.

The probability of the error is calculated as follows:

$$\begin{aligned} P = 2 \quad & \text{errors of delay or omission} \\ \div (8 \quad & \text{batches per day} \\ \times 340 \quad & \text{production days per year} \\ \times 20) \quad & \text{years of observation} \\ = 3.6 \times 10^{-5} \quad & \text{per check operation} \end{aligned}$$

This is one of the highest levels of reliability I have observed in execution of procedures, not including aircraft pilots.

As a curiosity, one of the technicians interchanged two read-only memory chips, one for the control computer and one for a safety computer, leading to the loss of a batch of resin. The probability for this error, in the absence of configuration control procedures and good labelling, is estimated to be about 50%.

Premature Termination of Production

In the same plant as the previous case, the product was drained down to the quench tank before proper cooling took place, resulting in rapid boil-up and large release of solvent vapour. For this case the error probability is estimated as follows:

$$
\begin{aligned}
P = 3 \quad & \text{errors of premature operation} \\
\div (8 \quad & \text{batches per day} \\
\times 340 \quad & \text{production days per year} \\
\times 5) \quad & \text{years for which press records were available} \\
= 2.2 \times 10^{-4} \quad & \text{per check operation}
\end{aligned}
$$

An Error of Omission

The treatment of flue gases from a chemical waste incinerator is important, to prevent the release of acid gases to the chimney. Acid gases are absorbed in a 'scrubber', in which milk of lime is sprayed into a chamber through which the flue gas flows. The lime reacts with sulphur dioxide to produce calcium sulphite and with hydrogen chloride to produce calcium chloride, both of which are absorbed in the milk of lime.

In this particular plant, the milk of lime was produced by pushing lime through a screw feeder into a water flow. The flow of milk of lime was monitored by a flow-meter, but the lime content, which should be 4%, could not be monitored (at least not easily and not in the actual installation). Any interruption in flow, such as a blockage by large lumps of lime, would lead to low water flow and a high lime ratio. If the lime content became too high, it could set fast like mortar in the spray piping. It took about 8 hours of work with cleaning rods and hydrochloric acid to clean the pipes, because the pipes had several bends (which in retrospect is a design error).

One operator overlooked the checking twice within 3 months. Since the checks should be once per hour, this represents the probability of an error of omission given as follows:

$$
\begin{aligned}
P = 2 \quad & \text{errors} \\
\div (3 \quad & \text{months} \\
\times 30 \quad & \text{days per month} \\
\times 24) \quad & \text{checks per day} \\
= 1.3 \times 10^{-2} \quad & \text{errors per operation}
\end{aligned}
$$

This probability is an average over three operators. The operator responsible for the two errors became unpopular (not just because of the difficult work of pipe and nozzle cleaning which the errors caused) and left to take a new job.

These first two cases represent almost three orders in magnitude difference in probability for two nominally identical error modes. The difference is certainly the degree of engagement and difference in the tendency to daydream on the part of the operators (as determined by observation in the control room). Note that the very high probability was regarded as unacceptable not only by the management but also by colleagues. A frequency of error of about one per year or higher is regarded as too high if it leads to excessive work or inconvenience.

Omission—Forgetting to Close Manifold Valves and Overlooking the Signs of Overflow

Multiproduct pipelines are used to transport various products such as gasoline, diesel and kerosene between product terminals. There is always some mixing of the products in the transmission, but the lines are so designed that this corresponds to 'minor contamination' when mixed in a large tank.

Multiproduct pipelines, and many other flexible ways of using tankage, require that there is a manifold pipe connected to all the tanks. Preparing for filling with a new product requires that the operator close all existing open valves and open the valve to the new tank. The timing for this needs to be quite precise to reduce the degree of product contamination. The speed cannot be all that high in many tank farms, however, because many of the valves are manual types.

If a valve is left open, the product may be pumped partially to the wrong tank. This leads to contamination and may lead to tank overflow if the level alarm fails.

There are several possible causes of omitting to close a valve. One can be closing the wrong valve, which is possible with a highly complex piping system, with poor labelling. Another possibility is distraction, for example, being asked by radio to perform another higher priority task first.

A third possibility is to assume that, for example, several valves are closed when in fact one or more are open; this can happen if the operator believes that the plant is in a certain state or if some unusual operation has been carried out of which he or she is unaware.

A fourth possibility is simple absentmindedness.

In one tank farm, operators (two, plus a supervisor) controlled an average of 10 product movements per day. There were two errors in one year, of relatively short duration (detected by unexpected level rise in the wrong tank in one case and by unexpected low filling rate in the right tank in the second case).

The error probability for this case is calculated as follows:

$$
\begin{aligned}
P = 2 &\quad \text{errors per year} \\
\div (10 &\quad \text{movements per day} \\
\times 365) &\quad \text{days per year} \\
= 5.7 \times 10^{-3} &\quad \text{per operation}
\end{aligned}
$$

The circumstances and causal factors here were the following:

- An experienced and knowledgeable supervisor
- A reasonably good control room display system
- No interlocking or valve lineup alarms
- Fully automated valving—no need for manual valve lineup

Errors of this type can be virtually eliminated by suitable automation. For example, a computer controller can give an alarm if there is more than one filling valve open at once.

A second example of this kind is that of an export line to a shiploading quay, from a gas plant. One line was used for the export of butane, propane and pentane. After the plant had been in operation for 16 years, closure of the pentane tank line was forgotten. Butane pumped to the waiting ship passed backwards into a 20,000 m^3 pentane tank. The butane, which has a boiling point and a temperature lower than those of pentane, evaporated almost instantly.

The tank roof blew off, and 20,000 barrels of pentane were lost (a truly frightening situation, which was dealt with satisfactorily by the emergency services).

The use of a single export pipeline for several products was an accident waiting to happen. After the accident separate pipelines were installed for each product.

Since the system had operated with a butane shipment every fourth day for 20 years, the probability of error must have been as follows:

$$
\begin{aligned}
P = 1 &\quad \text{occurrence} \\
\div (20 &\quad \text{years} \\
\times 91) &\quad \text{operations per year} \\
= 5.5 \times 10^{-3} &\quad \text{per operation}
\end{aligned}
$$

The circumstances for the incident were the following:

- The operator team not realising the possibility of reverse flow or the extent of the hazard
- A control room with limited automation and no feedback of valve position
- Manual valve lineup

Note that this plant did not have even the minimum protection in the form of check valves in the export line. Use of a common manifold/export line for two so incompatible fluids would today be regarded as a design error. In fact, the manifold was replaced by two separate export pipelines after the accident.

Omission—Overlooking a Level Alarm

While discussing valve lineup and the tank content, a similar type of problem may be mentioned. It is not necessary to forget a valve closure for accidents similar to those described above to occur. If the tank export valve fails to close, or leaks, then either the wrong product can flow into the manifold or the manifold product can leak backwards into the tank. Which alternative occurs depends on the relative pressures. Backward leakage can lead to contamination, or even to tank overflow. (This can occur even where contamination cannot, when all the tanks have the same content type.)

An operator cannot prevent this kind of effect from occurring, since the cause is random equipment failure. The effect of the leakage can often be observed, however, either by gradual increase in tank level or by an alarm. This kind of observation is a part of supervision of the plant and is discussed more extensively in Chapter 8.

The ability to observe such reverse flow depends on the kind of display in the control room. If all tank levels are displayed on a single screen, it is possible to track levels. If trend curves are shown, the effect is even more obvious; such displays are rare, however.

Provision of level alarms is more common, in fact more or less universal. A level alarm on a tank which is not involved in an inventory movement will always be surprising and difficult to diagnose. For the five cases of this kind in the incident database used for this report, none of the alarms were interpreted as legitimate. Four of the incidents led to overflow; the last was recovered incidentally.

For two tank farms, the frequency was estimated from control room incident logs:

$$\begin{aligned} P &= 22 \quad \text{occurrences} \\ &\div (11 \quad \text{years} \\ &\times 280) \quad \text{product transfers per year} \\ &= 7.1\times 10^{-3} \quad \text{per operation} \end{aligned}$$

Wrong Object—Confusion of the Item to Be Operated

The CSB reports an incident on a PVC reactor. An operator was to wash out a reactor, but the reactor drain valve was closed. The operator forced the valve to open by disconnecting a service air supply hose. Unfortunately, the reactor to which this operation was applied was the wrong reactor. It was in use and pressurised. Vinyl chloride monomer was released, ignited and exploded, and a large fire occurred [4].

This illustrates a relatively common occurrence, that of operating or maintaining the wrong item of equipment. Typical causes are the following:

- Poor labelling or complete lack of labelling of equipment
- Poor communication, in which the identity of the item is not identified or is identified ambiguously
- Erroneous identification on the person requesting the operation or maintenance
- Illogical physical ordering of equipment, e.g. with redundant pumps labelled in physical order as A, C and B

In 4.4 years of operation at another plant, two incidents occurred in which the wrong valve was closed. In both cases, this led to pump damage. Since the pumps were interchanged on a routine monthly basis, the probability of error is determined as follows:

$$\begin{aligned} P = 2 &\quad \text{occurrences} \\ \div (4 &\quad \text{years} \\ \times 129) &\quad \text{interchanges per year} \\ = 4.2 \times 10^{-2} &\quad \text{per operation} \end{aligned}$$

The circumstances for these events were the following:

- Experienced and well aware operators
- Poor labelling and arrangement

Wrong Object—Maintenance of the Wrong Item

It is possible to select the wrong item in maintenance as well as in operation. This is especially dangerous if the item is in operation.

For ammonia vessels, maintenance on the wrong vessel will lead to an ammonia release. If flanges are removed by loosening bolts and using a flange spreader to open the flange, the worst consequence will be a leak. If the maintenance procedure includes a pressure check or opening of a depressurisation valve, the release should not occur (or only if there is an additional equipment failure).

Opening a flange without using a careful procedure can leave the flange stuck until all the bolts are removed. Then the flange may be blown open by the pressure or may suddenly 'unstick' by displacement of the piping. Whether any of these effects occurs depends on the gasket material, the degree of tightening, the length of time the flange has been closed and the temperature.

Of 44 ammonia tanks operating over 10 years on average, one pressure safety valve was removed in error leading to an ammonia release. The probability of this is calculated as follows:

$$\begin{aligned} P = 1 &\quad \text{occurrence} \\ \div (44 &\quad \text{vessels} \\ \times 10) &\quad \text{years} \\ = 2.3 \times 10^{-2} &\quad \text{per vessel year} \end{aligned}$$

The conditions for this occurrence are as follows:

- Poor labelling
- Experienced maintenance artisans
- Limited familiarity with the plant in some cases

Too Late—Pump Draining

A maintenance technician opened a pump to drain to a sump, preparatory to removing the pump for maintenance. He then, contrary to instructions, went away for a break. The suction valve was not closed, so the contents of a tank drained down to the sump, which overflowed. The waste oil drained to the water-treatment plant, then to the sea. The technician returned too late to prevent the incident.

The plant in this case had operated for 12 years, with 10 pumps operating in this service. It was hard to determine how many times pumps had been maintained over this time, but the average over 2 years was three per year, all for seal or bearing replacement. The probability of the error is given by the following:

$$
\begin{aligned}
P = 1 &\quad \text{occurrence} \\
\div (12 &\quad \text{years} \\
\times 3) &\quad \text{maintenance tasks per year} \\
= 2.8 \times 10^{-2} &\quad \text{per operation}
\end{aligned}
$$

The conditions for this case were the following:

- An experienced but not hazard-aware technician
- A rather relaxed attitude to safety in the plant, later tightened up considerably

This case may be classified also as 'forgetting' to close the suction valve or poor communication in that the artisan assumed, without checking, that the pump had been isolated. (This is not an unreasonable assumption, since the pump should be isolated when taken out of operation, but the assumption should have been checked.)

Too Late—Tank Truck Filling

A tank truck driver connected up filling hoses to a gasoline tank truck and commenced filling. He then went to the amenities room to rest and smoke. The tank overfilled, and the gasoline flowed over and ignited. The truck loading rack was destroyed in the fire.

Prior to the accident, the terminal filled an average of 30 tank trucks per day (approximately; detailed records were unavailable). The installation had operated for 14 years. The probability of error is calculated to be as follows:

$$
\begin{aligned}
P = 1 &\quad \text{occurrence} \\
\div (30 &\quad \text{fillings per day} \\
\times 365 &\quad \text{days per year} \\
\times 14) &\quad \text{years} \\
= 3.8 \times 10^{-6} &\quad \text{per operation}
\end{aligned}
$$

This accident occurred before the advent of automated filling control for gasoline tankers. The very low frequency of errors reflects the awareness on the part of the truck drivers of the high degree of hazard and the high safety margin (15% of volume). Note that there may well have been more errors resulting in partial overfilling but which were recovered before overflow occurred.

The actual cause of the delay in responding could not be determined. Distraction by some other concerns is one possibility.

Too Fast—Rough Operation

There are several ways in which operators can start a system or perform an operation too fast. Examples are the following:

- Too fast filling of a pipeline by using excessive pumping pressure
- Too fast start-up of a plant before all checks are complete or all preparations are completed
- Too rapid heating of equipment (this can cause cracking of heat-sensitive vessels and excessive stress in compressors and can cause flanges to be loosened due to bolt stretching)
- 'Overrunning' of a distillation column with too high feed rate and boil-up (leading usually to poor quality product)

The reasons for too rapid operation are usually self-motivated pressure to produce or exceed quotas or management pressure to do the same. Self-motivation may be to achieve production targets or bonus thresholds or simply professional pride in effective operation.

In one reformer reactor, too fast start-up occurred in 2 out of 10 cases, giving

$$P = 0.2.$$

Circumstances were a fairly long period between start-ups (6 to 8 months) and a change in operators between the two cases.

Double Charging

Double charging of a reactor with liquid reactants occurred once in a period of 6 years, with loss of the batch. Production was on a 2-day cycle, so the probability of the error could be estimated as follows:

$$\begin{aligned} P = 1 \quad & \text{case of double charging} \\ \div (6 \quad & \text{years of operation} \\ \times 150) \; & \text{production runs per year} \\ = 1.1 \times 10^{-3} & \end{aligned}$$

Overlooking Low Suction Pressure

Low suction pressure on a pump causes cavitation over time. Low suction pressure can result from a blockage in the suction line or from a closed or only partially open valve. This occurred and was noticed on inspection on a small pump six times over a period of 4 years. The pump was activated every day except on weekends. (There might have been other occurrences, observed and corrected by the operator himself, but this seems unlikely.)

The probability of error is calculated as follows:

$$\begin{aligned} P &= 6 \quad \text{occurrences} \\ &\div 250 \ \text{operations} \\ &= 0.024 \end{aligned}$$

Factors affecting the probability were relatively untrained staff and the fact that the situation could be determined only visually by walking by or by careful listening.

Failure to Respond in Time to a Power Failure

The response to power failure described in Chapter 9 was successful. Four other plants affected by the same power outage could not recover in time, so that the plants shut down. The reaction time required for diagnosing the problem was in each case between 8 and 20 minutes (the time for level trips to stop the plants). Control systems in every case were powered by an uninterruptible power supply and continued operating properly.

The probability of error is as follows:

$$\begin{aligned} P &= 4 \quad \text{failures to respond in time} \\ &\div 5 \quad \text{plants} \\ &= 0.8 \quad \text{per response} \end{aligned}$$

Restarting Plant Operation with Tests Incomplete

In 11 major maintenance turnrounds carried out over a period of 20 years, only 2 carried out a complete programme of tests on emergency shutdown (ESD) valves, giving a probability of error of 0.8.

In another plant, testing and maintenance of solenoid valves was never carried out over a period of 20 years, because the valves were not listed on the original test and maintenance procedure.

Failure to Remove Trip Bypasses

In one boiler in a plant, a jumper wire installed during commissioning was left in place and caused a boiler explosion 22 years later when water level was suddenly reduced due to sudden excessive steam demand.

All other boilers in the plant (14) were inspected in the following investigation. No other bypasses were found. Since all must have had bypasses during initial testing during commissioning, the probability of error is estimated as $P = 0.067$.

In modern practice this error would probably be recovered, because level switches are largely being replaced by level transmitters, where the correct performance can be seen in the control room by comparing alarm and trip instrument readings. Note too that the accident is evidence that the level trip was not fully functionally tested over a period of 22 years.

Release Misjudgement

In pigging gas pipelines in a remote place, procedures stated that any condensed liquids cleared from the pipe should be drained from the pig receiver 'to a safe place'. In one case, operators interpreted this as to be to a small gulch, 60 metres from the site and downwind. The light liquids drained into a valley, but because the wind was very light, vapours spread over 15 kilometres down the valley before being ignited. Eleven persons were killed in the village where the ignition took place.

In this case, there was clearly a lack of knowledge to support the operator team's judgement. Clearly also, the standard procedure was unsafe. The procedure had been carried out an estimated 80 times before the accident occurred. So that

$$P = 0.0125$$

Opening a Filter

A filter was opened in a paint factory. The filter was blocked at the outlet and so retained pressure. The bleed valve was also blocked. When the filter was opened by uncoupling the quick-release toggle clamps, the cover flew off and hit the ceiling.

It is estimated from the marks on the ceiling that this had happened four times, the first three not reported. The number of filter cleaning operations was one per batch, with four batches per day over a period of 10 years, giving

$$P = 0.0012.$$

The circumstances were experienced operators, frequent need to clear blockages and a release mechanism that did not allow slow opening.

In another case, that of a very large filter on an ethylene cracker, the filter was opened while pressurised, giving a spurt of tarry residue, which hit one of the fitters performing the filter cleaning. Again, the filter vent valve was blocked. Filter cleaning occurred roughly once per month and the period of operation was 12 years, giving an occurrence probability of $P = 0.007$.

These cases are not ones of human error except to the extent of not observing the lack of pressure-venting flow and, in the second case, of not opening the filter cover bolts carefully. In interviews, many cases of vent blockage occurred, and the frequency of blockage was estimated about 1 in 10 cases, so the probability to take care, in a known latent hazard situation, was about $P = 0.01$. In the second case, the frequency of vent blockage was estimated as about 1 in 100 operations, so the probability of avoiding the latent hazard is estimated as 0.7, i.e. close to certainty. The difference in expectation of problems can explain the difference in frequencies.

Risk Taking in Construction

In two cases, chimneys were dropped during erection, due to poor slinging and error in judgement of the lifting problem. A total of 16 chimneys were erected by the company during the period studied, giving an error probability of about 0.12 per lift. This is a very high value, even for a difficult lift. The result indicates a fairly inexperienced team who did not know how to do the work and which took risks presumably in ignorance of the needs. From observations of the team on other lifts, this judgement was confirmed.

Other lifting accident frequencies for offshore work are given as 1.8×10^{-2} per crane year for dropped objects and 5×10^{-5} per lift [5]. This includes slinging errors, crane operation errors and defective equipment.

Commissioning Oversight

The piping to a ground tank was installed all the way to the top manhole flange. However, no hole was cut into the manhole cover. The pipe fitting was so precise that the lack of a hole could not be seen.

When the tank was commissioned, methanol was pumped to the tank. It sprayed out from the small gap between the end of the pipe and the manhole cover.

The pipe fitter had forgotten to inform that the installation was incomplete.

It is hard to judge the probability of error for this case. The number of opportunities for error is every pipe installed, but most such work is finished properly during the stage of mechanical completion and testing. In this case the pipe was not tested first, because this would have required filling water into the methanol tank. In all, the probability of leaving a pipe section untested after mechanical completion is estimated as about 1 in 300, based on this and other examples, so $P = 0.033$.

Driveaway without Decoupling a Hose, an Oxygen Plant and an Ammonia Plant

In an ammonia distributional terminal, a driver who had a heavily loaded schedule drove away without decoupling the filling hose. The pipe ruptured and released ammonia, but five separate ESD buttons were pressed by other drivers on the loading ramp and closed the ammonia supply valve. All the buttons were pressed within 3 seconds.

This was one case registered by the automatic system in the course of 5 years following installation of the recording system. Ammonia was delivered in this way for only about 6 weeks during the spring, with about 40 loads per day, giving an error probability of about $P = 8 \times 10^{-4}$.

Similar problems occurred on a liquid oxygen plan, with two oxygen releases resulting in the course of 1 week. The delivery of oxygen was about eight tank trucks per day; at the time of the accidents, the plant had been operating for 2 years, giving $P = 3 \times 10^{-4}$.

The circumstances for the incidents were relatively inexperienced drivers, an emphasis on meeting delivery schedules and no special protection devices. Driveaway accidents are so common that good design practice is now to provide ESD valves on the hoses, excess flow valves on trucks and breakaway valves on the piping. Best practice also involves interlocks using the truck starter key, booms or traffic lights interlocked to the hose lay-down point or the hose pressure.

Walking Away with Tank Draining

Water was to be drained from waste oil tanks prior to the oil being recovered and reprocessed. Draining off of water to the oil/water separator and wastewater-treatment system took quite a long time, and the operator left to smoke a cigarette in the smoking room. He took too long and oil flowed into the separator basin and then into the wastewater-treatment plant, contaminating it and putting it out of operation.

The water drain-off operation was a daily occurrence, and the error occurred after 5 years of plant operation. In this case, the plant operated 5 days per week and 48 weeks per year, so that $P = 8.3 \times 10^{-4}$.

This case shows the malleability of human error probabilities (HEPs). After the incident, the operator was dismissed, and no further occurrences occurred in the following 20 years.

Forklift Truck Driving Too Fast or Too Unstable Load

Forklift truck incidents leading to spills were registered for a company handling an average of 400,000 drums per year. The spill frequency was 1.3×10^{-5} per drum moved (typically 50 m around two corners) [6].

Piling on the Gas, Gasoline Transfer

Oil and product movements were made between a refinery and a port. Transfer operations took place at a rate of about 10 per day. This continued over a long period. On the first pipeline failure, due to corrosion, the pumping rate was increased, in order to keep up the pressure.

This error syndrome is so common that the probability must be on the order of 0.1 to 1.0 when ruptures arise and there is no alarm!

Erroneous Closure of a Drain Line

An operator drained a three-phase (oil, water and gas) separator to remove water. To speed the process and to avoid depressurisation and flaring, the draining down was performed pressurised, with a ‘cracked open’ valve. This could be regarded as a judgement error but not on the part of the operator, since this was standard practice.

At the end of the drain line, the wastewater plant operator found the API separator (oil–water, basin type) to be overflowing oil to the sea. He tried to, but could not, contact the plant operators, so he closed the drain line outlet valve. The drain line pressurised and ruptured, releasing oil into the plant. This was a judgement error, since the oil escaping into the plant was much more serious than the oil escaping into the sea. The cause was either lack of knowledge (of the pressurised draining) or lack of inference (of the effect of closure).

The probability of error is as follows:

$$
\begin{aligned}
P = 1 \quad & \text{event} \\
\div (4 \quad & \text{pressurised drain operations per year} \\
\times 14 \quad & \text{separators} \\
\times 22) \quad & \text{years} \\
= 0.000\,81 &
\end{aligned}
$$

Performance-Shaping Factors and Error Causes

In the early human reliability–analysis methods, baseline error probabilities were given which could be modified by means of factors to take account of causal influence. The classic archetype for such a performance-shaping factor is fatigue. There is no doubt that error probabilities increase with the degree of fatigue, as anyone can demonstrate for themselves by carrying out a standard task (such as adding a row of numbers or solving a Sudoku puzzle) when fresh and when tired. A baseline probability can be multiplied by a performance-shaping factor for fatigue if such an effect is relevant for the actual error scenario.

The performance-shaping factor approach is not used here for the following reasons:

1. If several performance-shaping factors are applied, the resulting human error probability (HEP) may become greater than 1.0.
2. Multiplying several factors together implies that the different factors interact, which is not necessarily the case. There is no evidence, for example, that the effect of fatigue is greater for inexperienced persons than for experienced ones. (In fact, there is some evidence that the effect of fatigue is less on younger and presumably therefore less experienced persons.) Multiplying performance-shaping factors implies that there is a strong positive correlation.
3. Logically, the effect of such factors as misinformation as an error cause is a simple yes or no, not one of a scaled effect.
4. Some of the influences of causal effects on error probability will be continuous (e.g. the degree of fatigue), and may not be linear.
5. Some of the causal influences for errors are completely independent causes, e.g. misinformation. Most of the causes of error considered in the action error–analysis (AEA) method are completely independent phenomena, and their contribution to error mode probability is additive, not multiplicative.

For these reasons, in this report, independent causes are treated separately, with their own probabilities. These probabilities for causes can be combined using the arithmetic of fault tree analysis (OR gates), which means that the probabilities will never be greater than 1.0.

True performance-shaping factors, which can affect the probability of a single error mechanism, may be necessary for some problem types such as fatigue. For these cases, it will be necessary to determine whether the error probability is continuously varying, depending on the degree of the causal factor, and whether the relationship between cause and probability is linear. The results of calculation of such separate causal influences can be combined using the fault tree approach.

SIMULATORS AS A SOURCE OF ERROR PROBABILITY DATA

For some kinds of data, even collections of cases over 35 years cannot yield enough results to give a reasonable coverage of the problems. Errors in responding to emergencies, for example, require that there are such emergencies. In any one plant, emergencies should occur less than once per year, making detailed data collection very difficult, verging on the practically impossible.

Good high-fidelity simulators, as used in nuclear power plant operator licensing and recertification, provide an opportunity to learn more, because the operators' responses to many emergencies are tested. Nevertheless, it takes many thousands of tests to yield data for a range of problems, when errors occur less than once in a thousand tests. Kirwan's studies give examples of the power of the technique [3].

One approach to this was to provide a general-purpose simulator originally developed to test operating procedures as they are written. It consists of an input system for process plant piping and instrumentation diagrams, a system for automated HAZOP analysis and a system for providing an operable HMI. It has been possible to derive human error data from this because of the efforts of operators and operation engineers in testing procedures.

SUMMARY OF ANCHOR POINT FREQUENCIES

Table 19.1 gives values for human error probabilities, based on values given above and in some cases from others. These sources were selected because they are based on original data, with no third-party judgement or reassessment involved.

The data in Table 19.1 are raw data taken from plant experience or simulator studies. The range of error modes and error causes covered is less than those observed in qualitative studies. The analyst will quite often meet situations where the data in Table 19.2 are inadequate for a particular case. Table 19.2 gives a derived set of values, selected to correspond to a 'normal' plant. Table 19.3 then gives a set of influences which may increase or decrease the probability of error. The normal condition is defined to have the following:

- Trained operators with at least some years of experience
- Well-trained supervisors, with many years or experience
- Written procedures, largely correct, but in any case not used by the operators because they are 'known by heart'
- Written procedures used directly for operations performed rarely (once per year or less)
- Close supervision of all new operators for at least 6 months

TABLE 19.1
Human Error Probabilities Derived from Experience

Activity Type	Error Description	Cause or Qualifier	HEP	EF	Source	No. of Cases
		Error Mode: Omission				
SOP execution	Omission of a step from execution of procedure	Procedure with check-off list, <10 steps	0.001	3	S&G	NA
		Procedure with check-off list, >10 steps	0.003	3	S&G	NA
		Procedure with check-off list, <10 steps, checklist incorrectly used	0.003	3	S&G	NA
		Procedure with check-off list, >10 steps, checklist incorrectly used	0.01	3	S&G	NA
	Omission of an item from a procedure using a tag list	Procedure with check-off list, <10 steps	0.001	3	S&G	NA
		Procedure with check-off list, >10 steps	0.003	3	S&G	NA
		Procedure with check-off list, <10 steps, checklist incorrectly used	0.003	3	S&G	NA
		Procedure with check-off list, >10 steps, checklist incorrectly used	0.01	3	S&G	NA
	Omission of an item from a procedure from memory	Well-trained and frequently used procedure	0.01	5	S&G	NA
		Less frequently used procedure	0.05	5	S&G	NA
	Omission of a procedural step at the control panel	Vitamin production procedure, 50 steps	0.003		TA	Many
		Restoration of a cooling system after testing, nitration reactor	0.0008		TA	1
		Forgetting to restore a valve position in a valve lineup, critical, pipeline centre with many operations per day	0.0057		TA	1
		Forgetting to restore a valve position in a valve lineup, noncritical, pipeline centre with many operations per day	0.046		TA	8
	Failure to close valve	Failure to manually close a valve at the end of 10-step procedure, isolated act at the end of a procedure	0.01		TA	7

(Continued)

TABLE 19.1 (CONTINUED)
Human Error Probabilities Derived from Experience

Activity Type	Error Description	Cause or Qualifier	HEP	EF	Source	No. of Cases
	Omission of a procedural step in the field	Forgetting to restore a valve position in a valve lineup, very critical, export terminal with 4 transfers per week	0.00089		TA	1
	Failure to carry out a frequent task step	Failure to check dosing of lime to milk of lime production	0.013		TA	2
	Failure to open suction valve	Failure to open the pump suction valve on a pump, inexperienced operator (2 months on plant, years in the industry)	0.04		TA	2
	Failure to start solvent supply	Failure to start solvent supply when starting an extractor, inexperienced operator	0.33		TA	1
COM	Omission of a procedural step in the field	Forgetting to remove an override jumper (bypass) from a level trip	0.067		TA	1
PM	Omission of test	Standard instrument testing routine	0.0013		TA	1
		Error Mode: Too Early				
UM	Premature opening of manhole cover	Manhole cover partially opened prior to completion of draining, high-stress situation	0.005		TA	1
SM	Premature restart of operations	Restart of operations before testing complete, under pressure to recommence production	0.18		TA	2
SOP execution	Premature draining down of product	Draining down of product to quench tank before properly cooled, under pressure of shift ending	0.00022		TA	3
	Premature driveaway	Tank truck driveaway before decoupling hoses, ammonia distribution	0.0003		TA	1
		Tank truck driveaway before decoupling hoses, liquefied oxygen distribution	0.0008		TA	1

(*Continued*)

TABLE 19.1 (CONTINUED)
Human Error Probabilities Derived from Experience

Activity Type	Error Description	Cause or Qualifier	HEP	EF	Source	No. of Cases
	Premature termination of task	Premature rundown of reaction products to quench tank	0.00022		TA	2
	Premature commencement of task	Admission of solvent to hot reactor without checking temperature, large puff of solvent vapour	0.00015		TA	2
		Error Mode: Too Late				
SOP execution	Too late return to task	Failure to get timing right for reactor, waiting period of 15 hours, distraction	0.000036		TA	1
		Failure to return in time to pump draining task, poor judgement, distraction, departure from work site forbidden	0.028		TA	1
		Failure to return from smoking room in time when filling tank truck, departure from work site forbidden	0.0000038		TA	1
		Late return from smoking room when filling tank truck, departure from work site forbidden	0.000065		TA	1
	Too late draining of condensate leg	Too late drainage of condensate from trap due to earlier valve difficulty, error in judgement of amount	0.0005		TA	1
		Error Mode: Too Fast				
Routine task	Driving too fast	Too fast driving of forklift truck, leading to overturning drum and liquid spill, per drum transport	0.000013		TA	1
Emergency task	Too rapid pump start	Too rapid fire water pump start in an emergency, risking water hammer damage	0.2		TA	1

(*Continued*)

TABLE 19.1 (CONTINUED)
Human Error Probabilities Derived from Experience

Activity Type	Error Description	Cause or Qualifier	HEP	EF	Source	No. of Cases
		Error Mode: Too Much Force				
UM	Overtightening of valve bolts	Tightening of valve bolts so much that valve broke, releasing high-pressure water jet and causing fatality	0.00029		TA	1
	Overtightening of valve or flange bolts	Tightening of valve or flange bolts using unauthorised tools and force	0.016		TA	8
SOP execution	Overtightening of valve	Overtightening of plastic sampling valve in plastic pipe so hard that it broke (together with a design error, too weak)	0.006		TA	1
	Excessive supply of reactant	Too many sacks of powder reactant supplied to reactor, poor system of counting sacks, distraction	0.05		TA	1
		Error Mode: Too High				
Routine task	Too many drums stacked	Too many drums stacked on top of each other, lack of knowledge, LTA training, LTA judgement, waste plant	0.002		TA	5
	Forbidden stacking of containers	Containers stacked on top of each other, lack of knowledge, LTA training, LTA judgement, pharmaceutical production	0.001		TA	1
		Error Mode: Repetition				
SOP execution	Double charging of reactor	Two charges of reactant supplied to kettle reactor	0.0011		TA	1

(Continued)

TABLE 19.1 (CONTINUED)
Human Error Probabilities Derived from Experience

Activity Type	Error Description	Cause or Qualifier	HEP	EF	Source	No. of Cases
		Error Mode: Forget–Remember				
SOP execution	Forget then remember supply of solvent	Supply of hexane to a soya oil extractor was forgotten by new operator, turned on by experienced foreman, release and explosion	0.0012		TA	1
	Forget then remember to start stirrer	Operator forgot to start stirrer on a bromination reactor, group of operators and supervisors decided to start it, knowing there was a risk; judgement error	0.00049		TA	1
		Error Mode: Wrong Direction				
SOP execution	Control turned in wrong direction	Equipment turned in wrong direction, equipment not in conflict with stereotype or norm	0.0001		TA	1
		Control knob turned in wrong direction, direction not intuitive or corresponding to norm, after instruction	0.05		TS	8
		Control knob turned in wrong direction, direction not intuitive or corresponding to norm, no instruction	1		TS	20
	Throttling valve turned in wrong direction	Control in field, adjustment of flow rate, experienced operators, relation between amount and response clear	0.07		TS	7
		Control in field, adjustment of flow rate, inexperienced operators, relation between amount and response clear	0.1		TS	2
		Control in field, adjustment of flow rate, inexperienced operators, response contrary to expectation	0.5		TS	5
	Rotary control turned in wrong direction	No stereotype violation	0.0005	3	S&G	4
		Stereotype violation	0.05	3	S&G	

(Continued)

TABLE 19.1 (CONTINUED)
Human Error Probabilities Derived from Experience

Activity Type	Error Description	Cause or Qualifier	HEP	EF	Source	No. of Cases
		Error Mode: Wrong Control				
SOP execution	Selection of wrong control	Wrong pump valve; dark conditions; poor labelling			TA	1
		Work station screen, controls arrange functionally	0.0004		TS	4
		Wrong tank selected to receive product, refinery	0.002		TA	1
	Inadvertent activation		Very low		S&G	3
	Selection of a control from an array	Identified by label	0.003	3	S&G	8
		Identified by functional grouping	0.001	3	S&G	2
		Part of mimic layout	0.0005	3	S&G	6
		Error Mode: Wrong Valve				
SOP execution	Selection of wrong valve to operate	Wrong valve out of two closed, identified by labels, dark conditions	0.042		TA	1
	Error in changing or restoring locally operated valve	Clear and unambiguously labelled valve, all valves similar	0.001	3	S&G	2
		Clear and unambiguously labelled valve, valves differ in one or more aspects	0.003	3	S&G	1
		Unclear or ambiguous valve labelling, valve set apart from others	0.0005	3	S&G	4
		Unclear or ambiguous valve labelling, valves differ	0.0005	3	S&G	2
		Unclear or ambiguous valve labelling, valves similar	0.01	3	S&G	3
PM	ECCS valves left misaligned after maintenance	Critical state for a highly critical system	0.0003		KA	1

(*Continued*)

TABLE 19.1 (CONTINUED)
Human Error Probabilities Derived from Experience

Activity Type	Error Description	Cause or Qualifier	HEP	EF	Source	No. of Cases
		Error Mode: Wrong Item				
Maintenance	Selection of wrong PSV	Wrong PSV from a pair dismantled or isolation valve in wrong position, ammonia	0.023		TA	1
		Wrong PSV from a pair dismantled, or isolation valve in wrong position, ammonia	0.021		TA	1
	Welders working on wrong pipe	Welders worked on the wrong pipe and created a hole	0.04		TA	1
SOP execution	Operator operated the wrong pump	Operator instructed to work on column 103, worked on 105	0.007		TA	1
	Wrong pump selected on mimic	Wrong pump selected on a mimic diagram, both pumps shown	0.008		TS	3
		Wrong pump selected on a workstation mimic diagram, pumps on different display screens	0.023		TS	7
		Error Mode: Wrong Value				
SOP execution	Invalid tag no entered into computer	Address for equipment to be manipulated had to be typed into a computer	0.005		TS	1
	Typing invalid set point values		0.002		TS	2
	Insufficient precision in setting pressure		0.005		TS	1
	Valves missed during calibration task	No feedback from system	0.001		TS	1

(Continued)

TABLE 19.1 (CONTINUED)
Human Error Probabilities Derived from Experience

Activity Type	Error Description	Cause or Qualifier	HEP	EF	Source	No. of Cases
	Wrong value in tank gauging	Mistake in reading of dip line	0.0003		TCRS	
	Wrong setting of rotary control	Two position switch	0.0001		S&G	
	Wrong setting of rotary adjustment	Rotary control with analogue scale	0.001		S&G	NA
	Failure to hold push button or rotary switch long enough	When there is feedback from the system	0.03		S&G	NA
		When there is no feedback from the system	0.003		S&G	NA
Supervisory	Wrong set point value entered to workstation	Temperature setting, disagreement between operators about operating goals	0.02		TA	4
		Operators tried to optimise production, larger amounts fed to kettle reactor	0.001		TA	2
	Wrong set point set for level trip	Operators wanted to speed draining down to surge vessel, left inadequate separation space in the space above the liquid, giving carry over to flare	0.03		TA	9
	Wrong set point value	Vitamin plant simulator scenarios, 4200 opportunities for error	0.00012		TS	5
	Wrong value selected on parameter displays identified by label	10 displays on one screen	0.009		TS	
	Wrong value selected on mimic diagram	10 displays on reactor and receiving tank display	0.0003		TS	
	Wrong value read on analogue-style display	10 set point/alarm/current value displays, error more than 10%	0.0003		TS	
	Wrong value read on trend display	10 trend displays on one screen, error more than 10%	0.011		TS	

(Continued)

TABLE 19.1 (CONTINUED)
Human Error Probabilities Derived from Experience

Activity Type	Error Description	Cause or Qualifier	HEP	EF	Source	No. of Cases
	Wrong value read on digital display	15 digital readouts on one mimic, error in 3 significant figures	0.0006		TS	
	Wrong value on field pressure gauge	Error more than 10%	0.02		TS	
	Wrong value reading level sight gauge	Mistaking level line	0.03		TS	
	Wrong value reading magnetic level gauge	Mistaking level reading or recording value wrongly	0.002		TS	
		Error Mode: Wrong Action				
Alarm response	Operators increased pumping rate when pipeline ruptured	Pumping gasoline to an export terminal, about 4 pipeline transport activities per day	0.1		TA	4
Emergency response	Operator shut drain line in response to separator overflow	Wastewater plant operator receiving too much oil into the API separator; he tried to telephone operators running the drain down but could not contact them; drain line ruptured, releasing oil, because of pressurisation; judgement error	0.00081		TA	1
Procedure execution	Operator recalling verbal instructions	1 item	0.001		S&G	NA
		2 items	0.003		S&G	NA
		3 items	0.01		S&G	NA
		Error Mode: No Response				
Emergency response	Operators failed to respond sufficiently quickly	In emergency exercise operators delayed average 1.5 min	0.25		TS	4

(Continued)

TABLE 19.1 (CONTINUED)
Human Error Probabilities Derived from Experience

Activity Type	Error Description	Cause or Qualifier	HEP	EF	Source	No. of Cases
		Error Mode: No Diagnosis				
Emergency response	Operators failed to respond sufficiently quickly	Operators in 5 out of 6 control rooms failed to respond quickly enough to a power dip	0.83		TA	1
	Operators failed to diagnose or did not respond	Simulation of instrument failure in 5 minutes	0.02		TS	4
		Simulation of temperature failure in 5 minutes	0.05		TS	10
	Vitamin production plant	Loss of cooling 5 minutes	0.06		TS	12
		Loss of cooling with trend display	0.005		TS	1
		Steam coil leak	0.08		TS	
		Pump leak, mass imbalance, 5 minutes	0.03		TS	6
		Error Mode: Delayed Diagnosis				
Abnormal operation	Commissioning engineers could not diagnose valve reversal	3-way valve installed reversed, allowing heated water to escape; diagnosis only after 45 minutes to drain	0.6		TA, TS	
		Error Mode: Planning Error				
Abnormal operation	Operating managers decided to restart with a faulty blower	Blower for a sulphur burner developed a heavy vibration and was shut down; after a planning conference, operation managers decided to restart cautiously; a blade flew off of the blower, destroyed the ducting and released sulphur dioxide; a lube oil fire started	0.5		TA	1

Note: COM: commissioning; ECCS: emergency core-cooling system; EF: uncertainty factor; HEP: human error probability; NA: not applicable; PM: planned maintenance; PSV: pressure safety valve; SOP: standard operation procedure; S&G: Swain and Guttman; TA: Taylor actual incidents; TS: Taylor simulator data; UM: unplanned maintenance.

TABLE 19.2
Generic Error Rates for Operations

Operation Type	Error Description	HEP
	Error Mode: Omission	
Standard operation, RB	Omission of a functional step in a procedure	0.001
	Omission of a check in a procedure	0.001
	Omission of an item from a long list (>10 items) from memory	0.01
	Omission of an item from a long list (>10 items) when using a checklist	0.003
	Isolated act, not related to main task function and separated in time	0.02
	Error Mode: Too Much/Too Little	
Standard operation, RB	Valve throttling when there is only auditory or direct observation feedback, no instrument	0.03
	Rotary control adjustment with feedback	0.005
	Slider adjustment on display screen with feedback	0.007
	Addition of solids or liquids, manual control or manual action	0.05
	Action when judgement is required	0.07
	Error Mode: Too Early	
Standard operation, RB	Premature operation in situation requiring waiting for completion or precondition to be fulfilled	0.0005
	Error Mode: Too Late	
Standard operation, RB	Too late action in situation requiring waiting for completion	0.03
	Error Mode: Too Fast	
Driving in plant	Driver drives so fast that overturning, drop of load or collision could occur	0.0001
Machine operation	Operator runs machine too quickly	0.001
Standard operation, RB	Valve opening	0.001
	Error Mode: Too Hard	
Maintenance, field operation	Too much force used in operating valve or other equipment	0.003
	Error Mode: Too Soft	
Maintenance, field operation	Operator unable to exert sufficient force	0.0 or 1.0, depending on equipment
	Error Mode: Wrong Item	
Field operations	General selection of wrong plant area or equipment item	0.0002

(*Continued*)

TABLE 19.2 (CONTINUED)
Generic Error Rates for Operations

Operation Type	Error Description	HEP
	Error Mode: Wrong Valve	
Field operations	Selection of wrong valve to operate when there are many valves	0.002
Switch panel	Selection of wrong valve to operate when there are many switches, structured	0.002
Mimic panel or mimic display screen	Selection of wrong valve to operate when well arranged	0.0005
Switch panel, functionally grouped	Selection of wrong valve to operate when there are many valves	0.002
Switch panel, not functionally grouped	Selection of wrong valve to operate when there are many valves	0.02
	Error Mode: Wrong Motor	
Field operations	Selection of wrong motor to operate when there are many motors	0.01
Switch panel	Selection of wrong motor to operate when there are many motors not functionally arranged	0.02
Mimic panel or mimic display screen	Selection of wrong motor to operate when motors well arranged	0.0005
Switch panel, functionally grouped	Selection of wrong motor to operate when there are many motors	0.002
Switch panel, not functionally grouped	Selection of wrong motor to operate when there are many motors poorly arranged	0.02
	Error Mode: Wrong Control	
Controller panel	Selection of wrong control to operate when there are many controls	0.0003
Mimic panel or mimic display screen	Selection of wrong control to operate when there are many controls	0.0002
Controller panel, functionally grouped	Selection of wrong control to operate when there are many controls	0.002
Switch panel, not functionally grouped	Selection of wrong control to operate when there are many controls	0.03
	Error Mode: Wrong Direction	
Valve in the field	Turning the valve the wrong way (if physically possible)	0.0001
	Error Mode: Wrong Value	
Digital input	Input on a workstation	0.01
Analogue reading	Input from control panel	0.003
Digital reading	Input from workstation	0.001

(*Continued*)

TABLE 19.2 (CONTINUED)
Generic Error Rates for Operations

Operation Type	Error Description	HEP
	Error Mode: No Response	
Supervisory	No response to indicator change, analogue display	0.01
	No response to indicator change requiring response, trend display, display continuously visible	0.001
	No response to annunciation with audible alarm change requiring response, digital display	0.001
	Alarm	0.001
	Error Mode: Delayed Response	
Emergency response	Failure to diagnose an alarm condition or to react to it, clear indication and specified response, single or few alarms	See error recovery below
	Failure to diagnose an alarm condition or to react to it, clear indication with priorities and specified response, many alarms (>5)	0.1
	Failure to diagnose an alarm condition or to react to it, no clear indication or specified response, single or few alarms (>5)	See error recovery below
	Error Mode: Incorrect Response	
Emergency response, supervisory	Operator misunderstands alarm or annunciation or responds incorrectly, on alarm panel	Depends on possibility for confusion; 0.3
	Operator misunderstands alarm or annunciation or responds incorrectly, on alarm workstation	Depends on possibility for confusion; 0.3
	Error Mode: Incorrect Diagnosis	
Emergency response	Incorrect diagnosis of complicated alarm, complex situation	Depends on complexity; 0.3
	Error Mode: Omission	
Standard operation, RB or emergency	Verbal communication forgotten or not heard, good auditive environment	0.0001
	Verbal communication forgotten or not heard, noisy environment	0.1
	Radio communication failed, field radio	0.03
	Message or information not written in log	0.1
	Message or information written in log not read or not understood	0.01

(Continued)

TABLE 19.2 (CONTINUED)
Generic Error Rates for Operations

Operation Type	Error Description	HEP
	Error Mode: Wrong Message	
Standard operation, RB or emergency	Verbal message misstated, ambiguous or misunderstood	Depends on practice; 0.01
	Error Mode: Erroneous Response to Message	
Standard operation, RB or emergency	Erroneous response to message	Depends on understanding; see above
	Error Mode: Delayed Response to Message	
Standard operation, RB or emergency	Delayed response to message, during emergency	0.3

Source: Taylor, J. R., A Catalogue of Human Error Analyses for Oil, Gas and Chemical Plants, itsa.dk, 2015.

Note: Data are also based on Table 19.1. Further data are given in Tables 2.1 and 2.2 of Ref. [3].

- Good equipment labelling
- All equipment logically arranged
- Consistency throughout the plant
- No special operating difficulties
- Checklists for long lists of items to be manipulated
- Equipment following good HMI practice, no serious deficiencies

USING THE HUMAN ERROR PROBABILITY DATA

When evaluating HEPs, it is easiest to evaluate influencing factors as a whole for the complete task first, for example, with the checklist given below. Many factors apply for a task as a whole.

When selecting data, the order of precedence should be as follows:

1. If there is experience from the actual plant, use it.
2. Select a value from an item which is closest to your actual situation in Table 19.1.
3. If there is no suitable entry in Table 19.1, select a value from Table 19.2.
4. Apply corrections for applicable special causes and influences by using Table 19.3 where applicable.

If there are causes which result in a large probability contribution so that the overall probability approaches 1.0, use an OR gate calculation as follows:

$$P_{\text{total}} = P_{\text{basic}} + P_2 + P_3 + \cdots - P_1 \cdot P_2 - P_1 \cdot P_3 - \cdots - P_2 \cdot P_3 - \cdots.$$

TABLE 19.3
Modifications for Special Error Causes and Influences

Cause or Influence	Error Modes Affected	HEP Calculation
Operator is unaware of the criticality of the task	All	Add 0.003 to the HEP.
Operator is inexperienced	All	Add 0.01 to the HEP.
Physical prevention, hindrance	All	Use the probability of hindrance occurring.
Distraction	Omission	Calculate the fraction of time that the operator is distracted (DT). Set DT to 0 if there are two operators. For an alarm, if the alarm has a blink and an audible alarm and individual resetting, add 0.01 × DT. If the alarms do not have individual resetting and there are several alarms, add 0.1 × DT to the HEP.
Distraction	Too late, delay	Calculate DT. Set DT to 0 if there are two operators. Add 0.1 × DT to the HEP.
Distraction	Omission, too late	Calculate DT. Set DT to 0 if there are two operators. When waiting for more than 5 minutes for completion of a process stage, add 0.02 × DT to the HEP. When waiting for more than 30 minutes for completion of a process stage, add 0.1 × DT to the HEP.
Poor access	All modes	Where there may be obstructions preventing operation, calculate the duration of the obstruction and the fraction of time obstructed (OT). Add OT to the HEP. Where obstruction can be overcome, add 0.1 to delay-mode or too late–mode HEPs.
Split job	Omission, too late	When the operator works at two different places, calculate the fraction of time he or she is away from the task concerned (AT). Add 0.1 × AT to the HEP.
Workload	All modes	When the workload exceeds the operator capacity, such as in an emergency, add 0.3 to the HEP.
Work stress	All modes	When work stress is high, such as in a critical restart where the plant is losing production, or there is a critical time limit for completion, add 0.01 to the HEP.
Personal stress	All modes	Where there is a possibility of personal stress, such as illness, family illness, marital problems or divorce, or money problems and support is poor, add 0.03. This will generally be difficult to assess in an analysis, so it is generally better to use this influence for sensitivity studies.

(Continued)

TABLE 19.3 (CONTINUED)
Modifications for Special Error Causes and Influences

Cause or Influence	Error Modes Affected	HEP Calculation
Lighting	All modes involving identification of pumps, valves, switches and controls	Where lighting will be poor, such as in an emergency response with lighting failure or a dark part of a plant such that reading is difficult, and there are several alternative items, add 0.3 to the HEP. Where there is only one item, add 0.03 to delay-mode or too late–mode HEPs.
Poor quality of labelling	Wrong item, etc.	Where labelling is poor, unreadable or missing, use the HEP for unlabelled items. In predictive analyses, it is better to recommend that good labelling is provided and to use this HEP value for sensitivity studies. If labels are missing or wrong, calculate the frequencies for opportunity for error, and use an HEP of 1.0.
Illogical ordering, e.g. tanks 1, 3 and 2	Wrong item selected, etc.	When ordering of switches, indicators or equipment is not logical, the situation is completely different to that of a well-designed plant. Use an appropriate HEP from Table 19.2 or add 0.1 to the HEP.
No functional grouping	Wrong indicator or switch selected	When functional grouping of indicators or switches is not made, the situation is completely different to that of a well-designed plant. Use an appropriate HEP from Table 19.1 or 19.2.
Number of items in list	Carrying out actions	Use an appropriate HEP from Table 19.1 or 19.2.
Knowledge, plant familiarity	All	Add 0.2 if the operator is generally familiar with operations but not the specific plant. Add 0.1 if the operator is a novice.
Knowledge, basic physics and needed chemistry	All	If physics or chemistry unknown to the operator is involved, use 1.0 as the HEP.
Knowledge, plant procedures	All	If the operator is unfamiliar with plant procedures, add 0.1 to the HEP.
Knowledge, hazards	All	If the operators have not had hazard awareness training, add 0.3 to the probability of error in response to the hazard.
Safety culture	All	If safety culture is poor add 0.02 to HEP
Systems bypassed	All	If safety systems are found to be bypassed for longer than the allowed interval, disregard the system and similar systems in error recovery.
Many alarms permanently red	All	Disregard alarms as an effective emergency response.

I.e. sum all the probabilities, then subtract the products of the probabilities for all pairs of cause contributions. This is the standard mathematical form for probabilities of events with multiple causes (simplified).

In reviewing the anchor point data in Table 19.1, it was found that there is a definite order in the impact of influences on probabilities. The most important ones are 'awareness of criticality of a task' and 'experience of more than 1 year'.

The following checklist is useful:

1. Is the task or activity critical?
2. Is the operator aware of the criticality?
3. Is the operator experienced in general with the plant and operations?
4. Is the operator experienced with the specific task?
5. Is the operator's job a split job, with activities at several locations?
6. Is there a possibility for physical hindrances in executing the task?
7. Is all necessary knowledge available? (This is difficult to answer, but you can get a good impression from feedback from hazard awareness courses or from HAZOP studies. Remember that if a person has been made aware because of HAZOP studies, etc., this is no guarantee that others on the team have been.)
8. Is the task
   a. routine, carried out on a daily basis or continuously?
   b. carried out periodically?
   c. carried out only occasionally or never before?
9. Is operation well disciplined, following procedures precisely and with only properly justified deviations?
10. Is there a general pressure for production?
11. Is the task stressful, requiring to be completed within a fixed time interval and with penalties or adverse consequences?
12. Are there distractions?
13. Are there competing priorities?
14. Are displays well structured?
15. Is the display for the task complex (more than 10 indications)?
16. Is the display in mimic form?
17. Is there good feedback from actions? E.g. are there position switches on valves which allow completion of actions to be displayed?
18. Is there an operating procedure for the task?
19. Is it up to date?
20. Is it used?
21. Is there a step-by-step checklist?
22. Is the equipment logically arranged?
23. Is there good labelling?
24. Is labelling weatherproof and maintenance proof?
25. Is the task dependent on instruments? If so, is there a history of false alarms?
26. Are there new and inexperienced operators?
27. Has there been training for operations?

28. Is the training material complete, including
    a. descriptions of unusual situations and the correct response;
    b. cautions and warnings;
    c. description of preparation and initial state and
    d. descriptions of required equipment and personal protective equipment?
29. Has the training material been checked for correctness?
30. Is any of the training material generic, and if so, is it completely applicable to and correct for the actual installation?
31. Can the task overload the operator, especially if inexperienced?
32. Is the safety culture good? Check this by considering operator awareness and use of correct procedures and by their reporting. Check also by means of mechanical integrity audit in the plant.

For individual task steps, the following need to be asked:

1. Is information for the step continuously displayed or in a lower-level display screen?
2. Is the step cued
    a. sequentially as a continuous sequence of steps?
    b. by alarm, feedback from the displays or annunciators?
    c. by direct visual observation?
    d. self-cued on the basis of passage of time?
3. Are there any physical limitations which affect performance?
4. Is the step isolated from the rest of the task and unrelated to the primary task function, e.g. restoration after testing?
5. Is there good access for the task step?
6. Can there be physical difficulties in carrying out the step?
7. Is there good feedback to indicate correct completion of the step?
8. Can consequences occur so fast that recovery is impossible?
9. Can the operator be injured or incapacitated if the step goes wrong?

ERROR RECOVERY

In some cases the feedback from the effects of an erroneous action is immediate and obvious. For example, if a liquid sample valve is opened to wide, the flow is large and if the operator or technician does not panic, the recovery can be immediate. For this kind of case, the probability of recovery is usually 1.0 unless the effect is dangerous or directly injures the person. I have observed five such situations, in all of which recovery was made quickly, but in each case this concerned experienced persons (see the example of driveaway incident in the previous anchor point examples). A conservative assumption, though, is to use a recovery probability of 0.5 if there are dangerous effects.

If there is good time for recovery, such as when liquids are directed to the wrong tank, the normal alarm and trip systems provide support for recovery. The probability of failure of recovery can be calculated by standard methods of safety system reliability analysis.

There are some situations in which an emergency situation exists but cannot be explained and cannot be immediately controlled by shutdown. Examples are given in Refs. [7,8].

Hannaman et al. [9], and more readily available in Ref. [10] developed curves relating probability of response to elapsed time from appearance of symptoms. It requires an estimate of nominal times to response from in plant observation simulator studies but gives the distribution of response times around the nominal value. Three curves are given, for skill-, rule- and knowledge-based actions. No validation of the curves for action outside the nuclear industry could be found.

VALIDATION

Validation of the calculations earlier is difficult because most of the actual data on error rates available have been used in deriving anchor point values for the error frequency calculations themselves. However, one data set was available. Over the course of years, the author has facilitated or carried out several HAZOP studies and AEAs for crude oil distillation units. These have a fairly standard design across the world. Validation is made possible by the British Petroleum (BP) safety booklet 'Hazards of Oil Refining Distillation Units' [11]. This gives 112 short accident descriptions and statistics on 144. With some additional effort, it was possible to bring the number of cases up to 160, with about 48 operator and maintenance errors. The other useful aspect of the data set is that the number of refineries in the world is known, so that the exposure can be determined. No data from Africa or Asia were used since the degree of accident reporting from these sources has been limited.

The accident prediction was made using an automated HAZOP study and AEA tool, which gives some degree of independence of the analysis results from the accident records used for validation. The data used for the human error prediction were independent of data used for the comparison, which is, of course, a negation of one of the first rules of quantitative analysis, that is, always use the most relevant data. However, this was necessary for a reasonable comparison.

A short summary of the comparison results is given in Table 19.4. More extensive data are given in Ref. [12].

TABLE 19.4
Validation of AEA Analyses against Historic Data

Operator or Maintenance Task	Ratio of Prediction to Observation
Pig receiving on oil pipelines	2.4:1
Changing a large bucket filter	1.8:1
Isolating a hydrogen sulphide absorber in preparation for maintenance	0.62:1
Replacing a motor for repair	3.1:1
Changing a heat exchanger tube bundle	7.7:1
Taking a sample of hydrocarbon condensate	1.6:1
Taking down a pressure safety valve for testing	0.87:1

REFERENCES

1. Marsh, *The 100 Largest Losses 1972–2011*, usa.marsh.com/Portals/9/Documents/100_Largest_Losses2011.pdf, Downloaded 2015.
2. O. Bruun, A. Rasmussen, and J. R. Taylor, *Cause Consequence Reporting for Accident Reduction: The Accident Anatomy Method*, Risø National Laboratory Report M-2206, 1979.
3. B. Kirwan, *A Guide to Practical Risk Assessment*. London: Taylor & Francis, 1994.
4. CSB, Vinyl Chloride Monomer Explosion, U.S. Chemical Safety Board, Report no. 2004-10-1-L, Washington, DC, 2007, www.CSB.gov.
5. J. Spouge, *A Guide to Quantitative Risk Assessment*, DNV Technica, 1999.
6. J. R. Taylor, *Hazardous Materials Release Frequencies for Process Plant*, 7th edition, 2006. www.efcog.org.
7. U.S. Environmental Protection Agency, *A Chemical Accident Investigation Report: Tosco Avon Refinery, Martinez, California*, 1998.
8. U.K. Health and Safety Executive, The Explosion and Fires at the Texaco Refinery, Milford Haven, 24th July 1994, http://www.hse.gov.uk.
9. G. W. Hannaman, A. J. Spurgin, and Y. D. Lukic, *Human Cognitive Reliability Model for PRA Analysis*, Draft Report NUS-4531, Electric Power and Research Institute Project RP2170-3, Electric Power and Research Institute, Palo Alto, California, 1984.
10. D. I. Gertman and H. S. Blackman, *Human Reliability and Safety Analysis Handbook*, Hoboken, NJ: Wiley InterScience, 1994.
11. British Petroleum, Hazardous Oil Refining Distillation Units. Institute of Chemical Engineers.
12. J. R. Taylor, A Catalogue of Human Error Analyses for Oil, Gas and Chemical Plant, http://www.itsa.dk, 2015.

20 Examples of Human Error Analysis

ACTION ERROR ANALYSIS FOR FUEL TRANSFER TO A TANK

An example of complete hazard identification is made here, to illustrate the principles from earlier chapters. The result is shown as a cause–consequence diagram in Figure 20.1. The identification, shown in Figure 20.1, is based on analysis of all steps in the procedure, using analysis sheets like those shown in Figure 18.1. Both possible human errors and latent hazards are identified.

The example chosen is a simple and often repeated one—transferring a liquid into a tank. Valve lineup is made so that mass transfer can be carried out in a plant.

Examples of such transfers are the following:

- Transfer of diesel fuel from storage to a day tank to allow a diesel generator to operate for at least 24 hours
- Transfer of enough solvent to a feed tank to allow a chemical batch reaction to proceed safely
- Transfer of crude oil from a ship to a storage tank in order to provide feed for a refinery
- Transfer of a liquid fuel from storage tanks to fuel tanks in a space satellite launcher
- Transfer of cooling water to an emergency cooling tank

The task requires the following:

- All valves along a particular flow route are opened.
- Any valves which would allow an unwanted flow to occur are closed.

For an experienced operator, the flow paths in the plant will usually be fairly clear, especially for those which are part of normal production. The operator may have a mental model of the actual physical flow, just a mental image of points along the flow or, in some cases, an image of the flow derived from mimic diagrams.

If a modern control system is used, there will generally be a mimic diagram to allow the flow paths to be established. In some cases, the mimic diagram will be spread across a number of different screens, in which case coordination will be more difficult.

For older control systems, the automated valve controls may consist of position switches on a board, requiring placement of the switches in the correct position. Switches may be grouped logically, but this is not always the case.

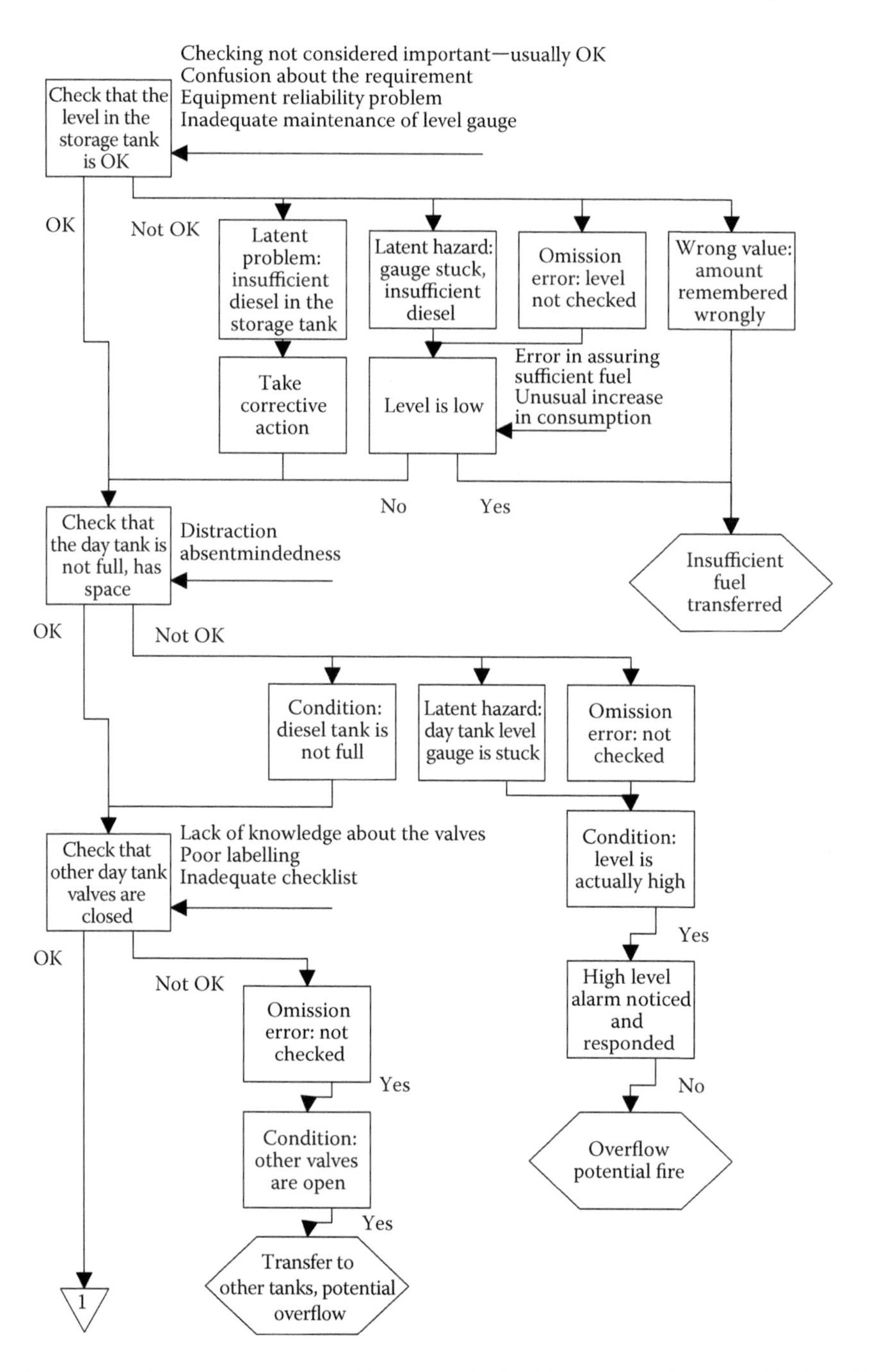

FIGURE 20.1 Cause–consequence diagram for the tank-filling operation. *(Continued)*

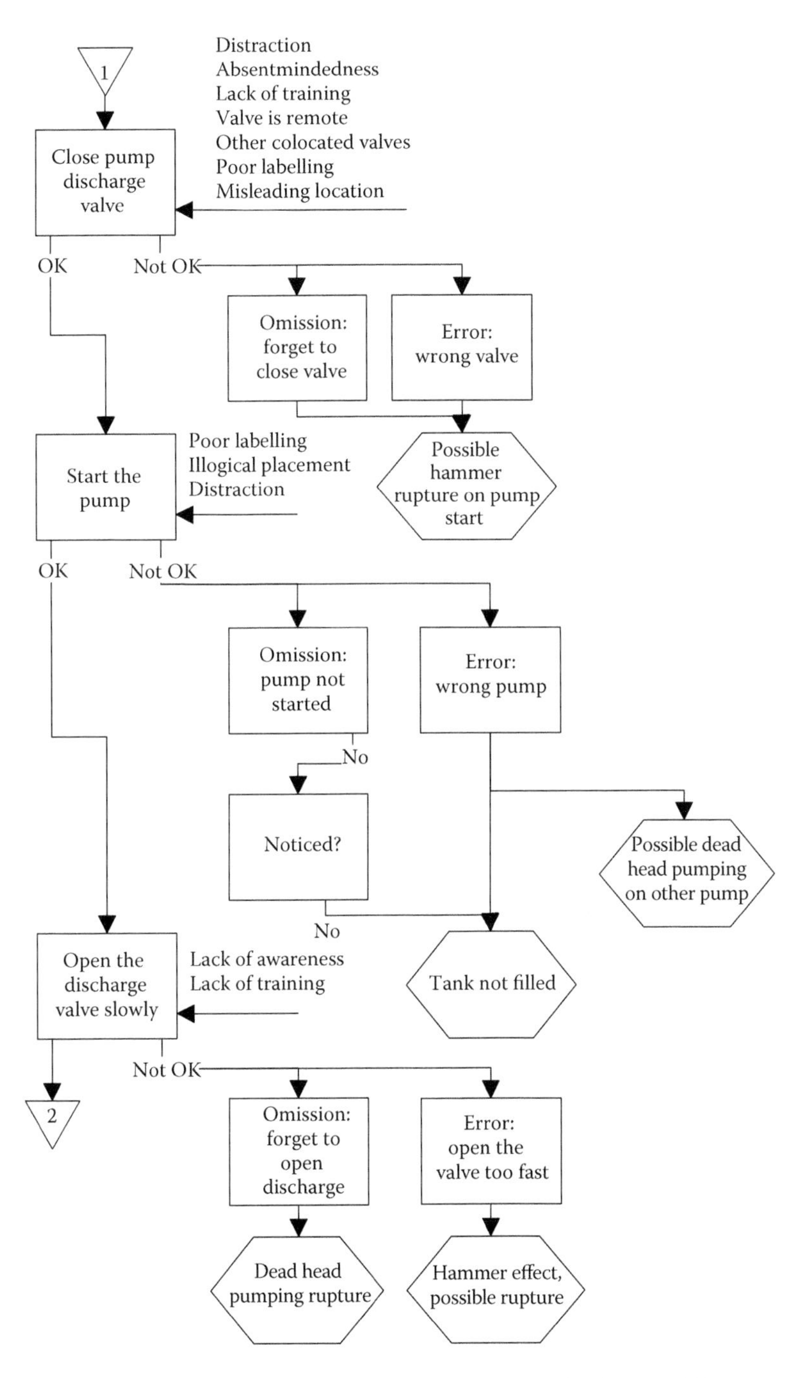

FIGURE 20.1 (CONTINUED) Cause–consequence diagram for the tank-filling operation.
(Continued)

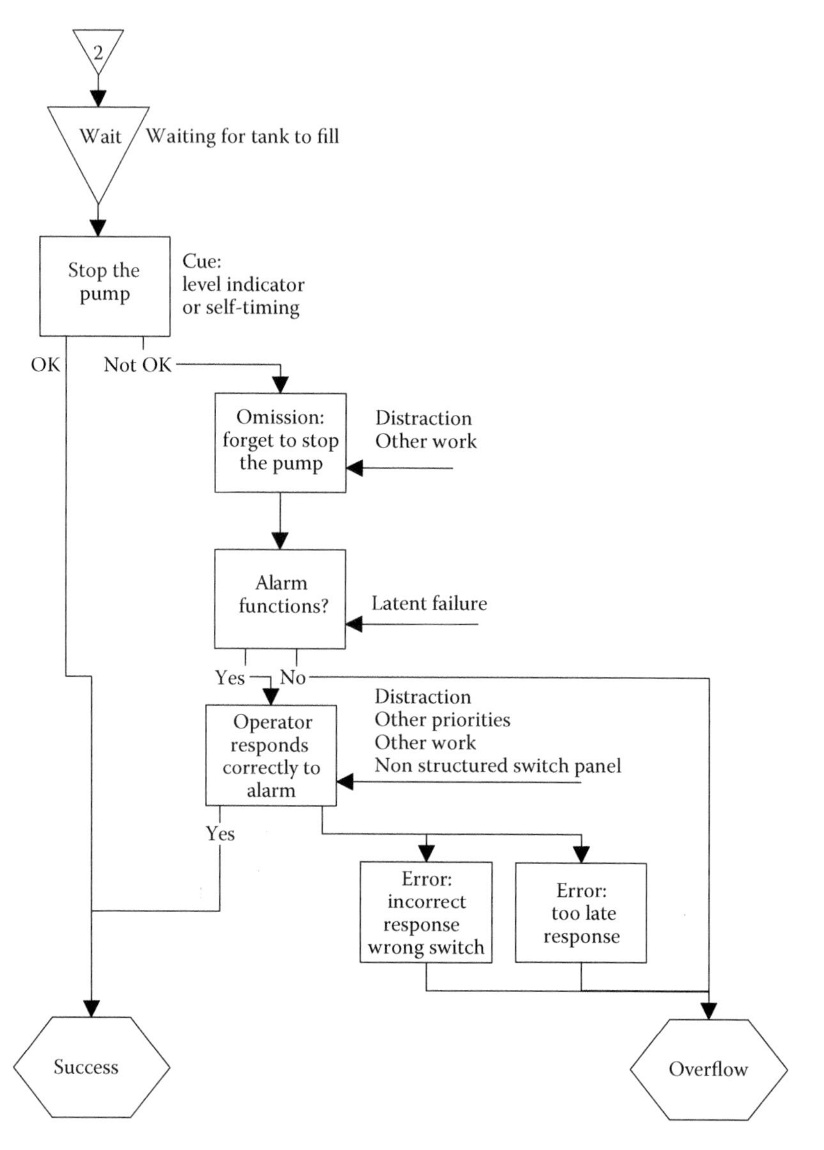

FIGURE 20.1 (CONTINUED) Cause–consequence diagram for the tank-filling operation.

In many cases, particularly those which involve procedures which are not part of everyday operations, such as draining off water from the diesel tank, the lineup may require that an operator goes out into the field and opens the necessary valves.

The procedure for transfer of diesel from a tank farm to a generator day tank has the following general form:

1. Check that the condition is correct for the transfer (if not, wait or correct).
2. Check that the existing valve lineup is correct.

3. If the valve lineup is not correct, close valves which are currently open and which would allow an unwanted flow.
4. Open valves along the intended flow path.
5. Start the transfer pump.
6. Check that the flow is properly established.
7. Monitor the flow until the day tank is filled.
8. Stop the pump.
9. Close the valves.

Step 1, checking that conditions are correct for the transfer, can involve such conditions as temperature, pressure and tank contents. Checking the existing valve lineup can involve checking an existing flow path or that valves such as drain valves are closed. Step 6, checking that flow is properly established, may involve, for example, checking that the flow rate is correct and checking that the pressure drop across a filter is acceptable.

Figure 20.1 gives a qualitative error analysis for the task, including possible latent hazards and error causes.

The possibilities of error include correct action in the presence of latent failures and hazards in the system, as well as operator error.

Step 2 of the procedure involves checking inventory. The inventory in the source tank should be sufficient to allow the goods of the transfer to be accomplished.

An error can be made as a result of a latent failure if the source tank level indicator is stuck or is reading wrongly. If the indicator is stuck, there will be a good chance for the operator to see the problem when transfer starts, because the level in the tank will not go down.

A simple error of omission will be possible for several reasons. The operator may make the omission by simply forgetting or by being distracted at the time of making the check. The check may be omitted because an instrument to be checked has failed, or because the operator assumes that it has failed. The check may also be omitted deliberately, because the operator thinks that it is not necessary.

Step 3 of the procedure involves checking that the existing valve lineup is correct. In most cases, this will simply involve checking that all valves are closed. The form of checking will depend on the type of display used. If a mimic diagram on a wall is used, valve position will be given by lamps. If a computer display is used, the state of valves may be given by their colour on the screen or, in some cases, by symbol shape as well. If a panel display is used, the state of the valves may be shown only by status lamps, rather than on a mimic display.

Possible errors in the checking steps are the following:

- The operator simply forgets to make a check.
- There is a latent failure; a valve which should not be open is open, and there is no real feedback.
- The operator mistakes the existing flow path and relates to another path.
- The operator mistakes or overlooks a valve on the flow path.
- The operator mistakes the status indication.

Simply forgetting to make the check can arise from a simple memory lapse or from distraction; if a mimic display is used, it will be very difficult to forget the check because the evidence will be directly in front of the operator. Necessarily so, because, otherwise, the operator cannot open the valves for the task. If the valve positions are given in nongraphic form, it will be easy to overlook the check or to make an incomplete check.

If valve status is not displayed at all, the operator may just rely on memory of the system state as it was left at the end of the last task or may rely on secondary indications, e.g. that there is no flow. In some rare cases, the operator might direct a field operator to make the check, but in practice, such checks are normally made as part of manual operation of valves.

TASK RISK ASSESSMENT

In preparing for maintenance or similar work, a method statement is usually required. This gives a description of the purpose of the job and the job steps. It should also give the tools and equipment required, the preparation needed, the safeguards to be in place and the protective equipment to be used.

A task risk assessment (TRA) is then required for most tasks. There is no standard methodology for TRA, but this usually involves defining the task, identifying the hazards and describing the safeguards in place. The risk assessment is then made usually using a risk matrix, in which the evaluation of risk is made in terms of likelihood of any accident, and the severity of consequences (see Figure 20.2). If the risk is found to be high, further safeguards must be selected. Residual risk should be low, and if this cannot be achieved, a justification is needed to explain why a medium level of risk should be accepted.

One of the problems in making TRAs is that the estimation of severity and likelihood of accidents is usually made by judgement or guesswork. Estimation of severity of accidents is usually not too difficult, although the severity of some kinds of accidents can surprise, e.g. the fact that hot water can kill [1]. More difficult is the judgement of probability or likelihood.

Since there is no standard methodology for TRA, the assessment methodologies actually used are often inadequate. A method which has been validated [2] is the following:

1. Define the scope of the task, the method statement on which it is based and the coverage of the task risk analysis.
2. Carry out a HAZID, for example, by using the methodology in International Organization for Standardization's Standard 17776 in order to identify the hazards present. The hazards should be tabulated and the precautions for each should be added.
3. An AEA, as in Chapter 18, should be undertaken, the possible errors and latent hazards tabulated, the consequences determined and the existing precautions added to the table.

Consequence				Frequency				
				A Less than once in 10,000 years	B Once in 10,000 to 1000 years	C Once in 10,000 to 1000 years	D Once in 1000 to 100 years	E Once in 100 to 10 years
Severity	People	Asset	Env.					
5. Catastrophic	Multiple fatalities or permanent disability	Extensive damage > $10 million	Massive effect long-term damage over a region	M	H	H	H	H
4. Severe	Single fatality or permanent disability	Major damage > $1 million	Major effect long-term damage over a large area	M	M	H	H	H
3. Critical	Major injury or health effects	Local damage > $100,000	Localised effect damage over a limited area	M	M	M	M	H
2. Marginal	Minor injury or health effects	Minor damage > $10,100	Minor effect short-term damage over a small area	L	M	M	M	M
1. Negligible	Slight injury or health effects	Slight damage > $1000	Slight effect small local spill	L	L	L	M	M

FIGURE 20.2 Typical risk matrix used for TRA. High risk must be reduced. Medium risk must be reduced if reasonably practical. L, low; M, medium; H, high.

4. A systematic lessons learned review is made, using case histories from earlier accidents in similar tasks.
5. Accident probabilities and consequence levels are assessed and then marked onto the risk matrix.
6. Additional precautions are added to bring all the risks into the low region of the matrix.

TRAs are time consuming to make and so are generally archived for reuse. This can lead to abuse, in which an old TRA is simply reviewed and the review process doing little more than confirm the old analysis and redefine responsibilities for providing safeguards.

A checklist which has proved useful in reviewing generic TRAs is the following:

1. Is the new task identical to the earlier one?
2. Is the size of the area affected the same?
3. Are there any new items of equipment, objects or critical lines such as power lines in the affected area?
4. Are the pressures, temperatures or toxic concentrations the same as for the earlier analysis?
5. Are there any new substances or materials which will be present?
6. Will new tools be used?
7. Will any new equipment be used, e.g. a new crane type or size?
8. Are the provisions for access for the new task just as good as for the old one?

TASK RISK ANALYSIS FOR EQUIPMENT ISOLATION

Equipment isolation is a critical task prior to maintenance on equipment. Many fatalities have occurred due to errors in installing isolations. The following is a task risk assessment made using the procedure described in Chapter 21. Results are given in Tables 20.1 and 20.2.

The method statement for the task gives the following steps:

1. The equipment to be isolated shall be clearly identified and clearly located in the plant.
2. The location of blinds shall be as indicated on the blinding drawing marked up and approved by the area engineer.
3. Gaskets to be used shall be as approved by the area engineer and shall be kept in bag or box until used.
4. The location of block valves to be used shall be identified with the assistance of a senior operator.
5. Location of the blinds shall be made with the assistance of a senior plant operator.
6. A green and a red tag shall be placed on the blind, marked with the tag number.
7. The area engineer shall keep a log of blinds installed.
8. Obtain a work permit for blinding.

TABLE 20.1
Steps of Task Risk Analysis for Equipment Isolation

Step	Hazard, Error	Consequence	Risk P	Risk S	Risk R	Existing Safeguards	Risk P	Risk S	Risk R	Additional Safeguards	Action by
Initial state	Toxic gas release (H_2S)	Possible fatality	C	4	H	SCBA, personal alarm	B	4	M	Area alarm	
			C	4	H	Area alarm	B	4	M	Confirmed safety distance to other groups	
	Flammable gas or liquid release	Possible fire, burn injury	C	4	H	Nonsparking tools, fire extinguisher	B	4	M	Fire watch	
	Heavy object—blind	Injury to hands or feet	C	3	M	Support derrick, rigging slings and eyebolts	B	3	M		
1. Identify equipment	Wrong equipment identified	Release of flammable gas under pressure on opening flange	B	5	H	Check equipment tag number, careful procedure for unbolting	A	5	M		
		Potential for large fire	B	4		Closure of block valves upstream and downstream		4			
2. Identify blind location	Wrong location	Possible entry of gas to the equipment when opened, possible fire or toxic injury	B	4		Cross-checking of tags on blind drawing, independent confirmation		4			
	Omission of one or more blinds	Possible entry of gas to the equipment when opened possible fire or toxic injury	B	4		Cross-checking of tags on blind drawing, independent confirmation		4			

(Continued)

TABLE 20.1 (CONTINUED)
Steps of Task Risk Analysis for Equipment Isolation

			Risk				Risk				
Step	**Hazard, Error**	**Consequence**	**P**	**S**	**R**	**Existing Safeguards**	**P**	**S**	**R**	**Additional Safeguards**	**Action by**
3. Obtain approved gasket	Wrong gasket type obtained	Possible leakage later, toxic injury	C	4	H	Area engineer approval	B	4	M		
	Reuse of old gaskets because new ones unavailable	Possible leakage later, toxic injury	B	4	M	Reuse forbidden	A	4	M		
4. Locate and close the block valves	Wrong block valve located	Release of flammable gas under pressure on opening flange	B	4		Careful procedure for unbolting, PPE	A	4	M		
	Block valve is passing	Release of flammable on opening flange	D	4	H	Careful procedure for unbolting, fire safeguards above, equipment nominally depressurised	B	4	M		
5. Locate the blind physically	Wrong blind located	Possible release of flammable under pressure on opening flange	B	4	M	Careful procedure for unbolting, PPE	A	4	M		
6. Place red and green tags on the blind	Omission	Blind may be left in place after restart	D	2	M	Area engineer or deputy checks tags in tag room	B	2	M		
7. Area engineer logs blinds	Omission—oversight	Blind may be left in place after restart	D	2	M	Cross-check between log and tag room	B	2	M		

(Continued)

TABLE 20.1 (CONTINUED)
Steps of Task Risk Analysis for Equipment Isolation

Step	Hazard, Error	Consequence	Risk P	Risk S	Risk R	Existing Safeguards	Risk P	Risk S	Risk R	Additional Safeguards	Action by
8. Obtain work permit	Premature operation, before PTW obtained	Possible interference with other activities	C	5	H	PTW to be displayed at the work site	B	5	H	Enforcement of PTW discipline	
		Possible toxic injury					A	5	M	Area engineer checks SIMOPs and cross informs all at nearby locations	
9. Ensure correct tools and spares	Wrong tools	Possible overtorqueing of bolts, crushing of gasket and leak	C	4	M	PTW officer checks tools	B	4	M		
		Possible ignition and fire, burn injury	C	4	M	PTW officer checks tools	B	4	M		
10. Ensure blind is ready and clean	Blind dirty	Possible leakage later, toxic injury	D	4	H	Clean the blind	B	4	M		
	Blind corroded	Possible leakage later, toxic injury	D	4	H	Confer with area engineer; if necessary, obtain new blind or refurbish	B	4	M	Provide blind condition guideline	
	Wrong blind	Possible leakage or rupture later, toxic injury	C	4	H		B	4	M	Check blind thickness and diameter; blind thickness to be specified in blind list	

(Continued)

TABLE 20.1 (CONTINUED)
Steps of Task Risk Analysis for Equipment Isolation

Step	Hazard, Error	Consequence	Risk P	Risk S	Risk R	Existing Safeguards	Risk P	Risk S	Risk R	Additional Safeguards	Action by
11. Obtain assurance from control room that equipment is depressurized and drained	Omission	Possible error in opening blind flange before depressurization or with much liquid	C	4	H	Careful procedure for unbolting, PPE	B	4	M		
	Pipe section not fully drainable	Possible release of liquid on blind opening	B	4	M	Careful procedure for unbolting, PPE	A	4	M	Arrangement for safe draining of residual liquid, spill tray and sand	
12. Use SCBA or air line equipment	SCBA worn but not in use	Possible toxic injury	C	4	H	Thorough H_2S awareness training	B	4	M		
	Air line kinks	Possible breathing difficulty, possible removal of mask	C	4	H	Air supply assistant present, responsible for air supply pressure and hoses	B	4	M		
13. Ensure block valves are closed	Omission	Extra safeguard is removed in case equipment is not fully depressurized or there are passing valves									
14. Support spacer with derrick or crane if heavy	Wrong method—manual support	Possible back injury or cut hands, foot injury	D	3		Derricks provided at most locations with large blinds	B	3		Provide guideline for maximum size of blind to be handled manually, need for crane to be specified on blind list	

(*Continued*)

TABLE 20.1 (CONTINUED)
Steps of Task Risk Analysis for Equipment Isolation

Step	Hazard, Error	Consequence	Risk P	Risk S	Risk R	Existing Safeguards	Risk P	Risk S	Risk R	Additional Safeguards	Action by
15. Loosen bottom bolts and ease the spacer; spread the flange and drain	Too fast, all the bolts loosened at once	Large release of liquid or gas if upstream or downstream valve is passing, toxic or burn injury	B	4	M	Earlier careful cracking open of the flange, PPE, fire precautions as above	A	4	M		
16. Allow to drain then remove all bolts	Problem if the gasket has been stuck and there is liquid in the pipe	Sudden release of liquid or gas on opening, possible toxic or burn injury	C	4	H	Earlier careful cracking open of the flange, PPE, fire precautions as above	B	4	M		
Coat with lubricant	Omission	Later bolt torqueing is inaccurate, possible leak of gasket	D	4	H	Good training, check that lubricant is available	B	4	M		
17. Install the blind using derrick or crane	Wrong method—manual support	Possible back injury or cut hands, foot injury	D	3	H	Derricks provided at most locations with large blinds	B	3	M	Provide guideline for maximum size of blind to be handled manually, need for crane to be specified on blind list	

(*Continued*)

TABLE 20.1 (CONTINUED)
Steps of Task Risk Analysis for Equipment Isolation

Step	Hazard, Error	Consequence	Risk			Existing Safeguards	Risk			Additional Safeguards	Action by
			P	S	R		P	S	R		
18. Insert gasket on both sides of the blind	Gasket becomes damaged or dirty	Possible leakage later, toxic injury	C	4	H		B	4	M	Quality control and training for gasket installation; this is especially important later when removing the blind and fitting the spacer or closing the flange	
19. Tighten all bolts	Too much force	Possible crushing of the gasket, leakage later	C	4	H	Bolt torqueing procedure, approved tools	A	4	M		
20. Remove the red tag; leave the green	Forget	Confusion about blinding status		1		Area engineer follow-up		1			
21. Close the work permit	Omission, too late	Problems in progressing other tasks		1				1			

Note: H: high; L: low; M: medium; P: probability; R: risk; S: severity; SIMOPs: simultaneous operations, see Figure 20.2.

TABLE 20.2
Earlier Occurrences

Cause	Consequence	Reference
Crane supported blind flange under tension, blind sprang up when the bolts were removed	Large release of liquid	U.K. Health and Safety Executive report [3]
Premature spade removal	Gas release causing toxic injury to nearby workers	YouTube video [4]

9. Ensure that correct size of spanners, other tools and spare bolts, nuts, studs and washers are available.
10. Ensure that the blind is ready to be inserted and is clean and in good condition.
11. Obtain assurance from the area engineer that the equipment is emptied and depressurized.
12. Personal protective equipment (PPE) must be worn and used during blinding, including self-contained breathing apparatus (SCBA) or trolley-supplied breathing air.
13. Ensure that upstream block valves (and downstream block valves if fitted) are closed.
14. If the blind spacer is a large one, ensure that it is supported by a derrick or by a crane if necessary.
15. Loosen the bottom bolts for the flange and spread the flange slightly to allow any liquid or gas escapes. Note that all the bolts will need to be loosened slightly, the bottom bolts more than the others.
16. When no further liquid or gas escapes, remove the remaining bolts, and clean them; coat them with lubricant.
17. Install the blind, using a derrick or a crane if necessary. Use a flange spreader if the space between flanges does not allow the blind to be inserted freely.
18. Insert gaskets on both sides of the blind and locate precisely by fitting three bottom bolts to finger tightness.
19. Tighten all bolts by using standard tightening procedure that uses torque wrench or calibrated hydraulic wrench.
20. Remove the red tag and store in the tag room.
21. Close the work permit.

REFERENCES

1. J. R. Taylor, Industrial Accident Physics, http://www.itsa.dk, 2014.
2. J. R. Taylor, A Catalogue of Human Error Analyses for Oil, Gas and Chemical Plant, http://www.itsa.dk, 2015.
3. UK HSE. The Fires and Explosion at O Oil (Grangemouth) Refinery Ltd. 1987. UK Health and Safety Executive (undated).
4. GASCO, Near Death H_2S incident. www.Youtube.com/watch?v=z/7VTtCpw9E, downloaded Sept. 2015.

21 Human Error Risk Reduction

The measures tabulated in this chapter do not represent a full set of those which constitute a full safety management programme. They are summary names for groups of safety management methods which can prevent major hazard accidents. Each of the following sections discusses a safety management measure.

OPERATOR ERROR MINIMISATION

Human error frequencies can be reduced by the following:

- Making procedures logical and explaining why individual steps are needed
- Making sequences of equipments logical
- Standardising operation approaches across an entire company
- Design for ease of operation and inherent safety

TRAINING FOR RELIABLE AND SAFE OPERATION

Good training is a prime route to safer operation, especially if it leads to operators understanding what is happening inside the equipment. Training should not be just in how to carry out the procedures. Training is needed for the general principles of equipment operation, about the actual plant, about safety and about organisation. Understanding also requires knowledge of the physics and sometimes the chemistry involved in the plant. A typical gap is in limited teaching of hazard awareness and the underlying physics of accidents, which can be very important in preventing or responding to accidents.

GOOD DISPLAYS

Modern operator workstation design is quite advanced, with good access to overall displays and a hierarchy of unit displays, process parameter control displays and trend displays. The area under control is generally monitored by fire and gas detectors, and modern designs show these in map form.

In modern plants, operators will generally have control over a large number of CCTV cameras and may be able to display several locations at once. A recent innovation is coupling of cameras to fire and gas detectors, so that cameras automatically focus in on trouble spots.

ALARMS AND TRIPS

Alarms, trips and interlocks, coupled with shutdown and depressurisation valves and pump trips, are the prime defences against accident consequences in a plant. In modern practice process, parameters of pressure, temperature and level will all be monitored by alarms and trips if these parameters are critical for safety, as they mostly are. In some places, other parameters such as flow may be monitored by alarms, especially if these are included in leak-detection systems. Trips and the ESD system are almost always required to be independent of the plant control system for safety's sake.

Alarms are intended to allow operators to take action before parameters exceed the limits of the operational envelope. Operators can do this if disturbances develop slowly. Response in the form of manual intervention for even a simple alarm can take 1 or 2 minutes and can take much longer if investigation is needed.

It is important to realise that not all accident types can be detected by alarms and that many occur too fast for successful operator response. Leaks of hazardous material may be detected by gas and vapour detectors and alarm, but area coverage of these is seldom above 80%, and there is a significant response delay, typically about 30 seconds after gas or vapour has reached the alarm.

There are many types of plant disturbance which cannot be instrumented either. Levels in stirred reactors and viscosity in resins are parameters which can be important for safety, but for which instrumentation is difficult.

Accidents such as pipe leaks and breakage generally happen too quickly to allow for good operator response. It may take 1 to 2 minutes to confirm that an accident is actually taking place unless it can be seen directly on CCTV screens and unless several alarms are activated in a clearly recognisable pattern. Hannaman et al. [1], among others, investigated time reliability curves, which give the time for a reliable response to alarms. The work has been mostly on nuclear power plants. Typical times for reliable alarm response on these plants were found to be 1 to 10 minutes, but for some alarms on these systems, response time was observed to be hours. Such long response times were found during the studies for this book to be very unusual for petroleum plants, which have a simpler causal structure, but a 10-minute response is not unusual.

Much research has gone into investigating the capacity of operators to respond to alarms [2]. A good estimate which seems to apply across a wide range of plants is 10 alarms per hour as a maximum. This has led to considerable effort to improve alarm systems. This includes alarm minimisation workshops. Changes include the following:

- Elimination of nuisance alarms which do not serve any function.
- Conditional alarms, so that alarms no longer appear when not relevant or when they are expected: Low-temperature alarms during start-up, before heating has taken effect, for example, are suppressed.
- Filtering of repeat alarms or chattering: This used to be handled by requiring parameters to go significantly beneath the alarm threshold at least once before an acknowledged alarm was reactivated. More sophisticated filtering is now used.
- Groups of related alarms are collected under one alarm signal, avoiding showers or floods of alarms.

In a modern plant, if the operator fails to control a disturbance within the available grace time, and if the disturbance is dangerous, the plant should shut down automatically, on ESD. This does not necessarily stop the accident, but it should limit the extent. Battery limit ESD valves on plant units prevent a continuous flow of feed into the unit and allow depressurisation to work, but there is usually a considerable inventory of hazardous materials in feed vessels, columns, and accumulators. Inventory isolation ESD valves on these vessels limit hazardous material releases to the content of the piping. Selection of these, though, is usually dependent on the results of risk analyses.

There may be disturbances in a plant which are not usually protected by trips. Overflow of liquids into flare lines is one example. Full protection of plant up to modern standards is very dependent on the quality of HAZOP studies.

INTERLOCKS (PERMISSIVES)

Interlocks are devices, either mechanical or electronic, which are intended to prevent actions being carried out unless it is safe to do so. A typical interlock on a batch reactor will prevent opening of a valve to admit solvent until the reactor is cool enough to avoid immediate boiling or flashing and release of vapour from the reactor loading manhole. Another typical interlock prevents admitting very active reagents such as acid to a reactor until it is established that the stirrer is functioning. In oil and gas pipelines' scraper receivers, a standard mechanical interlock is on which prevents opening of the receiver door until the receiver itself is successfully depressurised.

Selection of interlocks is easy, systematic and thorough if based on the AEA method described in Chapter 18. In fact, the method described there was developed originally in order to facilitate interlock design [3].

PERMIT TO WORK SYSTEMS

A PTW provides a way in which hazards and safety precautions can be independently checked before work starts. PTWs do not usually apply to plant operation but always apply to maintenance and to construction activities.

PTWs usually involve large forms. The supervisor or engineer organising the work usually fills out the first part of the form giving the location of the work and the type of work to be done. The permit officer will then fill out a standard checklist of safeguards which are necessary, including personal protective equipment. Different types of permits are usually given, for hot work, cold work, vessel entry, etc. The officer will also specify any tests to be carried out such as the presence of breathable air in any vessels or tanks to be entered. The permits are usually prepared a day before the actual work to be done, sometimes earlier. A critical part of the work permitting is work site inspection, where a permit officer checks that the appropriate safeguards are in place and that there are no unrecognised hazards.

An area engineer will then review the permit and may sign off, so that work can proceed. The area engineer will need to take into account other activities in progress which may be incompatible with the new task. It is usual for the permit office to keep a map of activities, which is updated on a daily basis. A matrix of permitted operations is usually used also to describe which activities are compatible. For example, radiography can usually be carried out when no other activity is taking place and when all persons not involved are well outside the area affected by radiation. Also, heavy lifting is generally not allowed above an operating plant, so crane locations and lift paths may be marked on plot plans.

Permits may be issued for many days, which reduces the effort in making them, but they must be renewed every day or sometimes at every shift, to ensure that no new hazards have arisen.

PTW systems have themselves been subject to error, particularly with signing off the permit without actually inspecting the work site. Also, the actual safety distances to be allowed around a work site are often unclear or incorrectly specified. A common safety distance required around a hot work (welding, flame cutting, shot blasting, grinding, etc.) is 15 m and there should be no flammable materials within this distance. Such distances may not take into account the possibility of flammable vapour migrating to the work site or collecting at low points. Also, Carborundum cutting disks can throw sparks up to 30 m, and sparks from welding and flame cutting can travel further if the work is being carried out at height.

Attachments are often required to be appended to work permits. Typical permits are heavy lift analyses and risk assessments, the method statement and task risk analysis.

FIGURE 21.1 Example of a control room designed for tall people only.

HUMAN FACTORS

Human factors is a term covering a wide range of studies. It includes ergonomics of comfort and well-being at work, ergonomics of access to equipment and controls (see Figures 21.1 through 21.3), ergonomics of access and design of controls for easy understanding and interpretation. Sometimes human factors is also considered to cover human error analysis. This can lead to problems, because as described above, human error analysis needs to be made in a multidisciplinary way. Human factors

FIGURE 21.2 Valve controls that are very difficult to operate.

FIGURE 21.3 Valves designed for optimum convenience and effectiveness.

specialists can and should take part in safety analyses rather than providing separate human factors reports to cover human error.

Refs. [4–7] provide good guides to the modern thinking on design of the human–machine interface for error reduction. Ref. [8] gives a good introduction to ergonomics as applied in industrial plant. Refs. [9] and [10] give good guides to alarm system design, especially for avoiding large numbers of alarms and alarm avalanches.

REFERENCES

1. G. W. Hannaman, A. J. Spurgin, and Y. D. Lukic, *Human Cognitive Reliability Model for PRA Analysis*, Draft Report NUS-4531, Electric Power and Research Institute Project RP2170-3, Electric Power and Research Institute, Palo Alto, California, 1984.
2. Abnormal Situation Management Consortium, Effective Alarm Management Practices, Abnormal Situation Management Consortium, http://www.asmconsortium.com, 2008.
3. J. R. Taylor, *Interlock Design Using Fault Tree and Cause Consequence Analysis*, Risø National Laboratory Report M-1890, 1976.
4. IOGP Human Factors – A means of improving HSE Performance. International Association of Oil and Gas Producers, 2005.
5. IOGP Human Factors – IOGP Human Factors Engineering in Projects, 2011.
6. M. S. Sanders and E. J. McCormick, *Human Factors in Engineering and Design*, McGraw-Hill, 1992.
7. EEMUA, Publication 191, Alarm Systems – A Guide to Design, Management and Procurement.
8. C. Sandom and R. S. Harvey, ed., *Human Factors for Engineers*, London: The Institution of Engineering and Technology, 2004.
9. IEC 62682, Management of Alarm Systems for the Process Industries, International Electrochemical Commission.
10. ASM Consortium Effective Alarm Management Guidelines, ASM, Honeywell, http://www.asmconsortium.org.

22 Conclusion

This book describes one approach to human error analysis, the action error analysis method and its underpinnings. The method is known to be effective, since it has been used in over 100 analyses. The method was validated qualitatively in a demonstration project between 1978 and 1982 [1], with follow-up of the plant on which it was used over several years. The approach to human error identification has also been validated in aerospace and weapon system safety applications, so there may be a wider area of application beyond that of oil, gas and chemical plants.

Over a period of 35 years it has been possible to collect enough data to allow the method to be used in quantitative analysis of human error probabilities, an area which is now much more important than in the 1980s because of the present-day heavy reliance on risk analysis and the ALARP principle in safety engineering.

It has been possible to validate to a certain extent the quantitative aspects of human error analysis. To do this, a comparison was made between the predictions made using the method in the 1990s and the actual accident frequencies since then. There are just a few plants where this is possible. It proved impossible to validate the original accident frequency calculations, because the data documented in this book were not available at that time. It was possible to substitute the new data back into the old calculations and to derive new frequencies. The results of the comparisons are given in Ref. [2].

Alongside the development of the operator and maintenance analysis method, it has been possible to observe deeper, into management error and design error. Whether these can be predicted, and the frequency calculated, is doubtful. Whether they can be prevented is certain. All it takes is careful thought, good intentions, a lot of hard work and usually also a reasonable budget.

One theme that runs through much of this book is the importance of knowledge, and the importance of lack of knowledge. This knowledge covers a wide range, from knowing the physics underlying explosion phenomena, and therefore how to prevent explosions, to knowing how individual bolts need to be tightened in order to ensure plant safety. Some of the knowledge is arcane, such as the mechanisms of the more special types of corrosion. I am often amazed, how small causes can have huge effects. Much of this knowledge can, at present, only be acquired by experience. The next steps in safety are almost certainly to make that knowledge more readily available, so that it is less necessary to wait for the accident, before learning how to prevent it [3].

REFERENCES

1. J. R. Taylor, O. Hansen, C. Jensen, O. F. Jacobsen, M. Justesen, and S. Kjærgaard, *Risk Analysis of a Distillation Unit*, Risø National Laboratory Report M-2319, 1982.
2. J. R. Taylor, *A Catalogue of Human Error Analyses for Oil, Gas and Chemical Plant*, http://www.itsa.dk, 2015.
3. J. R. Taylor, *Industrial Accident Physics*, to be published.

Bibliography

The methods described in this book represent an approach more or less frozen in time at about 1978, with a fairly long programme of data collection and validation following this. There are many other approaches to human error analysis and many other good books which give an alternative view. The ones which have been most useful to me are listed below.

1. J. Reason, *Human Error*, Cambridge, UK: Cambridge University Press, 1991.

   This book gives a good overview of the subject of human error and its related cognitive psychology as it stood in the early 1990s. A good part of the terminology for human error analysis was established in this book. It describes the generic error-modelling system (GEMS) model, which provides additional detail to the skills–rules–knowledge model of Rasmussen.
2. T. A. Kletz, *An Engineer's View of Human Error*, Second ed., London: The Institution of Chemical Engineers, 1991.

   This is a short book, crammed with examples of actual human error incidents, organised according to ways of reducing error. As the title states, it contains a good set of examples to stimulate safety engineering.
3. D. I. Gertman and H. S. Blackman, *Human Reliability and Safety Analysis Handbook*, Hoboken, NJ: Wiley InterScience, 1994.

   This is a well-organised collection of the methods used for operator error analysis in the nuclear industry.
4. J. Reason, *Managing the Risks of Organisational Accidents*, Farnham, UK: Ashgate, 1997.

   This is a very readable description of management and organisational error, which goes well beyond the rather focussed description in this book. It should be read by anyone trying to organise an improved safety management system or to assess organisational safety performance.
5. S. Dekker, *Drift into Failure*, Farnham, UK: Ashgate, 2011.

   This is a systematic description of the ways in which organisations can push their operations beyond safe limits. It contains a strong criticism of the use of simple causal models in evaluating the background to accidents and their use in assessing failures in complex systems.
6. A. J. Spurgin, *Human Reliability Analysis*, Boca Raton, FL: CRC Press, 2010.

   This book provides a well-organised, up-to-date survey of human reliability assessment methods and human error data collections.
7. B. Kirwan, *A Guide to Practical Human Reliability Assessment*, Boca Raton, FL: CRC Press, 1994.

   This is an extensive survey of human reliability–analysis methods up to 1996, and it is special in that the methods are compared and validated against a suite of standard analysis problems, with examples from the chemical industry, nuclear power industry, oil well drilling and offshore oil production.

8. E. Hollnagel, *Cognitive Reliability and Error Analysis Methodology – CREAM*, Amsterdam: Elsevier, 1998.

   This book describes in a very clear way the further development of the cognitive models in human reliability analysis.

The methods described in the present book are based directly on the work of Jens Rasmussen. To review all of his work, and the many collaborative projects in which he engaged, would require a book in itself. There are three major works, though, which are important for the further development of human reliability analysis, even though not directly focussed on this.

9. J. Rasmussen, A. M. Pejtersen, and L. Goodstein, *Cognitive Systems Engineering*, Hoboken, NJ: Wiley InterScience, 1994.

   This book is intended to provide a link between cognitive psychology and engineering of systems. It provides an update of the skills–rules–knowledge model and its application in work analysis, design and work organisation. An extended example is given of cognitive modelling applied to the design of a library system.

10. J. Rasmussen, K. Duncan, and J. Leplat. *The Definition of Human Error and a Taxonomy for Technical System Design*, Chichester, UK: Wiley, 1987.

    This gives a definitive human error taxonomy.

11. J. Rasmussen and K. Vincente, Coping with Human Error Through System Design: Implications for Ecological Interface Design, *Int. J Man-Machine Studies*, (1989), 31, 517–534.

    This paper is based directly on the SKR model and show a way in which it can be used in the design of the human machine interface.

Index

C

D

E

F

G

H

I

N

O

P

Q

R

S

T